THE
SEKEM
EFFECT

Translated by Jeff Martin

Photograph on p. 34 courtesy of Right Livelihood.
All other photographs by Samuel Knaus and Justus Harm.

First published in German as
Sekem Inspirationen: Impulse für einen zukunftsfähigen Wandel
by Info3 Verlagsgesellschaft Brüll & Heisterkamp KG, Frankfurt in 2022
First published in English by Floris Books, Edinburgh in 2025
© 2022 Info3 Verlagsgesellschaft Brüll & Heisterkamp KG
English version © 2025 Floris Books

e Also available as an eBook

Authorised EU Representative: EasyAccess System Europe,
Mustamae tee SO, 10621 Tallinn, Estonia
gpsr.requests@easproject.com
British Library CIP data available
ISBN 978-178250-943-1

THE
SEKEM
EFFECT

How a Sustainable Community
Can Transform Egypt and the World

HELMY ABOULEISH
with Christine Arlt

Floris
Books

Contents

Preface   7

Part 1: The Sekem Initiative   13

1. Introducing Sekem   15

2. Sources of Inspiration   18

3. The Sekem Symphony: From Initial Vision to the Present   28

Part 2: The Sekem Vision Goals   37

4. An Overview of the Sekem Vision Goals   39

5. Culture   43
   Vision Goal 1: Lifelong learning   60
   Vision Goal 2: Holistic research   64
   Vision Goal 3: Integrative health   67
   Vision Goal 4: Arts and culture   70

6. Ecology   74
   Vision Goal 5: Agriculture   85
   Vision Goal 6: Sustainable water management   90
   Vision Goal 7: Renewable energy   93
   Vision Goal 8: Biodiversity   97
   Vision Goal 9: Climate neutrality   100

7. The Economy   104
   Vision Goal 10: Circular economy   115
   Vision Goal 11: The economy of love   118
   Vision Goal 12: Ethical banking and finance   122
   Vision Goal 13: Sustainable lifestyle   125

8. Social Life   129
   Vision Goal 14: Agile organisational structures   140
   Vision Goal 15: Diversity and equal opportunities   143
   Vision Goal 16: Social transformation   146

9. Our Vision for Egypt 2057 in the Context of Current Events 150

Conclusion   157

The Sekem Future Council Members (2025)   159

Acknowledgements   161

Notes   163

Resources   166

Index   169

Preface

We have a vision. A vision of the future of Egypt. It shines towards us and fills us with joy and a thirst for action.

When I say 'we', I mean the Sekem Future Council. Currently, this council consists of sixteen people, all of whom have made a conscious decision to share in the responsibility of implementing the Sekem vision. My father, Ibrahim Abouleish, described this role as follows:

> One of the tasks of this council is to maintain a living
> connection to the well of spiritual inspiration. A further task is
> for the leading members to experience the connection to others
> as an enrichment and completion. Social ability also entails that
> every individual has an awareness of all the others, that they
> know the conditions of the others and which tasks they are
> working on ... Through discussions with the outer world and
> the attitude that there is a solution for every problem, people
> can grow and work together. They become able to stand up
> consciously for the development of the people and the world.[1]

There is no better way to describe the commitment of my Future Council colleagues. In this book, we want to tell you about our inspirations, visions and deeds.

Time and again, we are struck by how negative the view of the future is. Yet despite all the stumbling blocks, there are so many great solutions that can create a positive future. One reason for this confidence is that we do not just imagine a future shaped from the past, but also look at the potential that comes to us from the future. This approach is the legacy of my father, who formulated the Sekem vision in 1977 with this awareness.

The landscape provided inspiration for the Sekem vision.

Forty years later, my father passed away. He left us his vision and, with the Sekem initiative, a multifaceted entity that has grown out of it and encompasses a wide variety of activities. Through this development, some questions for the future have become clear to us:

- How can my father's vision work into the future?
- How will the Sekem community evolve without its founder?
- How can the vision remain alive in the Sekem initiative?

My father left without any worries, because he had been preparing me to succeed him for the past forty years. When the time finally came to take on this responsibility, it became clear to me and my colleagues on the Future Council that it was up to us to find answers to these questions.

My father was a charismatic character who, as a founder and visionary, had insights into and an awareness of Sekem's development and purpose like no one else. For me and my colleagues, it is not really a question of succession in the sense of replacing my father, but rather a question of further development and change. One person cannot and should not replace the founder; Sekem should be shaped by the spirit of many. For this purpose, my father had already founded the Future Council. This council must now develop from a group of committed people who helped shape Sekem into an organised body that can carry the vision forward in a living and dynamic manner. We are currently on this rocky and bumpy, but always exciting and promising, journey of discovery.

The Sekem initiative has many parts to it that always aim to work holistically. This means that in each sphere of activity, the other spheres are included as a complement. At the same time, however, each sphere remains free and independent of the others. To place this task, which was led by our founder for over forty years, in several hands requires a complex restructuring process that takes time. Each one of us brings our own individuality, and the more space this receives, the more effective the collaboration becomes.

Two generations: Dr Ibrahim Abouleish and Helmy Abouleish.

Through the Sekem initiative the desert landscape blossomed.

Looking back, I can see that this process began for me back in 2011, when I spent a hundred days in prison. During the revolution in Egypt, I, like quite a few other entrepreneurs, and also ministers and activists, had been arrested on charges of corruption. I was released after three months and the charges were later dropped. I used this time to reflect on a number of issues that I had never found time and leisure for in my overcrowded workday, among them, my inner ability to recognise how important it is to face the future with an open and receptive mind. I also dealt with my relationship to Islam, which brought me back to my spiritual sources, and I reflected on my role in Sekem and my goals for the future, as well as on my task in Egyptian society. This began a process that led me away from the national and international platforms that had preoccupied me before, and towards a focus on Sekem. The importance of continuous individual inner training for the further development of Sekem became clearer again. From my point of view, with such a comprehensive vision, we can make much more of a difference in the world than just through politics.

It is our intention in this book to show such a vision and take the next step by encouraging many pioneering projects that support a systemic change in Egypt. I am convinced this will contribute to a successful future for our country, principally due to the experience gained after the founding of Sekem and during the pioneering stages, when we had to prove that our vision was indeed viable and that our doubters were wrong.

We can draw on the practical experience of forty-five years of implementing our vision. We are rooted in spiritual sources of inspiration that carry us safely, and we have people working with us who are capable of questioning and developing themselves anew every day. On this basis, it is now our intention to create and share practical solution models with the rest of Egypt, and potentially beyond.

In this book we want to share this process with you. We want to report on the forward-looking projects we have started, how we are shaping the future while remaining open to new ideas, and why we are working so confidently and joyfully in this area!

Helmy Abouleish

Part I:
The Sekem Initiative

I.

Introducing Sekem

The Sekem initiative is known to many people today as a 'miracle in the desert'. Some speak of a flourishing farm in the middle of the desert, others of holistic educational institutions for the rural population of Egypt. Still others know our companies for organic products or have heard of a sustainable community of Christians and Muslims from all over the world. What Sekem is and how its principles are lived cannot be described in a few words. Perhaps that is what makes Sekem so unique: its versatility in wholeness, its continuity despite constant change and development. Even those who have visited Sekem once and have experienced its complex structure will most likely find new projects, new people and new facilities when they return. Forty-five years after its founding, Sekem is indeed a phenomenon: the vision of a holistic initiative for sustainable development put into practice. A place where the new is embraced – often with many compromises and sometimes hard-to-bear ambiguities – and then actively developed, often in ways quite different from what was initially thought. And yet, at the same time, Sekem is far from being perfect. There is still so much more to be done.

This book sets forth the origins of Sekem, how the four fundamental spheres of the cultural, the ecological, the economic and the social work together, and how the Sekem vision intends to grow into the future. This journey has been both utopian and intensely practical from the beginning, for the Sekem vision is far from complete. After forty-five years of practical experience, concrete solutions for sustainable change can be multiplied. What is expected of a conventional company,

Starting the morning together. One of the many daily circles at the Sekem Farm.

namely growth, is not a central concern for us. Rather, it is a matter of intensifying what has been created and taking up incoming impulses in such a way that models emerge that can be adapted and widely disseminated, models that contribute to a real change in existing structures.

First we will give a brief overview of the historical development of Sekem and its sources of inspiration. Then we will look more closely at the four different spheres of impact: the cultural, the ecological, the economic and the social spheres. Finally, we will share our vision of 'Egypt 2057' and our sixteen Vision Goals. Within the framework of these sixteen goals, we have already launched quite a few projects. Some have already begun to have an effect on Egyptian society, others are still in the research and development phase. The sixteen Vision Goals are intended to be models of solutions for the challenges we face today and which will enable us to realise our vision of a sustainable future for Egypt.

These challenges tell us that the world needs to undergo a change of consciousness, or rather that a transformation of consciousness is currently underway. The Sekem vision sensitises us to this new consciousness and to the different stages of consciousness that are at work in the world. We recognise more clearly that in contemporary times many procedures are characterised by a consciousness that is predominantly focused on the intellect and on matter; art, culture, emotions and spirituality are neglected or reduced to their material aspect. Convinced as we are that long-term sustainable progress is not possible without acknowledging emotions and the spiritual dimension, we additionally want to address the topic of consciousness and its development in this book.

We have chosen to give an extensive presentation of the sources of inspiration to illustrate the importance of having a holistic view of things as well as a comprehensive approach to development.

2.

Sources of Inspiration

I am often asked about the spiritual background of Sekem.
Sekem developed out of my own vision. My spiritual inspiration
came out of very different cultures: a synthesis between the
Islamic world and European spirituality. I moved around freely
in these different areas as if in a great garden, picking the fruits
of the different trees.

Ibrahim Abouleish[1]

The genesis and implementation of the Sekem vision – that is, the
realisation of the Sekem initiative – is closely linked to Ibrahim
Abouleish's biography and his sources of inspiration. Throughout his
life, he was able to inwardly connect with diverse currents and impulses,
from the East and the West, Egypt and Europe, Islam and Christianity,
spirituality and science, thus creating a synergy of inspirations:

I existed in two worlds, both of which I felt were completely
different from each other: the oriental spiritual stream I was
born into and the European, which I felt was my chosen
course. But during this time I also started experiencing
moments when these two streams met in my soul, when I
was neither Egyptian nor European … The two completely
differing worlds within me slowly began to dissolve and
merge into a third entity, so that I was neither completely
the one nor the other. I could live in both worlds, think both

ways. But what I experienced was not a cheap compromise, nor just tolerance, but a synthesis, even an elevation in the Goethean sense, a real uniting of the two cultures within me ... I had become something else, and that is how I wanted to remain – even concerning my religion ... I wanted to achieve this state of being a 'third' in religion too, to be able to live within both and through this transcend to a higher level of being.[2]

Thus, the inspiration for the Sekem vision came from a wide variety of sources: from Islam as well as from anthroposophy; from Ibrahim Abouleish's native Egyptian culture as well as from the European one; from Goethe and Beethoven, as well as from the Arab mystics. Ibrahim Abouleish had always shared these diverse impulses in various ways, whether it was during the regular meetings of the Sekem co-workers, or with people who came to him from all over the world to learn about his interpretation of the Quran and its practical implementation. He approached things courageously and with a great enthusiasm for action, constantly changing his approach. This caused more than a few people to despair, but he also amazed them by immediately conceding when their objections seemed to serve the cause. The Sekem initiative, with its various fields of activity, is a living example of this approach. We like to refer to a saying by Rudolf Steiner, the founder of anthroposophy:

Seek the truly practical material life, but seek it such that it does not numb you to the Spirit working in it. Seek the Spirit, but seek it not in supersensible lust, out of supersensible egotism; seek it because you want to become selfless in practical life, selfless in the material world. Turn to the old maxim: never Spirit without matter, never matter without Spirit! Do this so that you can say, 'We want to perform all material deeds in the light of the Spirit, and we want to seek the light of the Spirit in such a way that it develops warmth within us for our practical deeds.'[3]

In this chapter we will describe the inspirations – from Islam especially, but also from art and literature across East and West – that have guided the Sekem vision and its practical implementation. Ibrahim Abouleish wrote: 'For the vast majority in the world Islam is a mystery in its meaning and its striving. We hope this background becomes more apparent through our work.'[4] As far as the interpretations of the sources of inspiration are concerned, we do not claim to be exhaustive. Rather, we wish to describe the interpretations we have applied in Sekem – efforts that we consider more important today than ever and that have guided us to our vision of the future. It is important to note that the quotations from the Quran that we cite are translations and interpretations made by Ibrahim Abouleish himself, based on Adel Theodor Khoury's translation of the Quran.

We feel it is important to present the sources of inspiration to show how the sphere of 'worldly' activity connects with the spiritual sphere, something that hardly receives any attention in modern times. Experience has taught us that the connection of practical life with the spiritual is a prerequisite for real, long-term sustainability. Ibrahim Abouleish was convinced that our vision could not be realised without a daily, concrete connection with the spiritual world.

In this chapter, we will also describe how we transform historical spiritual insights to make them relevant to the present, and how they shed light on major current issues. The old wisdoms do not lose their topicality, and we are often astonished when urgent questions of today appear in earlier times.

Developmental impulses from East and West

At the heart of the Sekem vision is development: a holistic, sustainable development that includes culture, nature, the economy and society all working in balance to serve the development of humanity and the world. This central idea is what connects the aforementioned sources of inspiration.

The first revelation of the Prophet Muhammad, Surah 96, can be understood as a call for development. The archangel Gabriel

People on their way to work and study in Sekem.

commands the prophet: 'Read!' For the Prophet Muhammad, who is illiterate, this command refers more to the search for knowledge and the confrontation with his own self and the world rather than any actual reading of the written word. In this surah the ability to know is attributed to human beings. At the same time it is pointed out how this cognitive faculty can lead to self-importance and arrogance, which prevents human beings from having a unified experience of the world. In our current technical age we often experience this self-aggrandisement in the scientific worldview, which pushes spirituality aside or denies it altogether. Science and technology do not exclude spirituality, however. Rather, it is the complete detachment of a subject from its living context that makes science imperfect.

Following the instruction to read, Gabriel then asks three questions:

Have you recognised the one in you who prevents the human being from connecting with the divine? Do you know the one who leads on the straight path, the one who makes God present? Have you recognised the one who denies the truth and turns away? (Surah 96:9–13)

These three questions are intended to test human beings and indicate the possibility that exists for them to freely carry out their own development. Question one speaks of that temptation by which human beings believe themselves to be divine and not in need of divine grace. Question three points at the danger of denying truth or the divine and acknowledging only matter. But it is the middle question that points to the path of development that humanity can take. When and how they accomplish this development is placed in the hands of each individual, as the Quran emphasises:

Before him and behind him he has angels, who guard him at Allah's command. Allah does not change the condition of people, until they themselves change their condition.
(Surah 13:11)

We see a task for the future in making it possible for every member of our community to freely choose his or her path: to be firmly anchored on the earth and yet find access to the spiritual, to the divine, to a realm other than the purely material.

In addition to the individual task of development, however, humanity has also been given responsibility for the world:

We offered the responsibility for the world to the heavens and the earth and the mountains, but they refused to carry it, they shrank from it. The human being bore it. (Surah 33:72)

When the Quran subsequently adds, 'He is really foolish!' or 'He overestimated himself out of ignorance,' this can be understood as an indication that we human beings do not live up to this responsibility – the issues are numerous and well known. No human being wants to consciously do harm, yet nevertheless it happens. If we are to bear this world responsibility, then we must acquire the necessary awareness for it. Sekem would like to contribute to this task.

Ibrahim Abouleish found access to this central idea of development in the works of Goethe. The German poet and naturalist was initially the reason why the young Ibrahim, at the age of nineteen, could not be dissuaded from his resolution to travel from Egypt to Europe to study. The work of Goethe aroused in him a longing to learn about European culture, in which he found further sources of inspiration. Goethe's contemplation of nature, his holistically oriented methodology, his understanding of Islam, but also the theme of development, which appears again and again in his work, inspired Ibrahim Abouleish throughout his life. He did not see human nature as being unchangeable, but as something that relied on continuous development. He particularly enjoyed quoting from Goethe's poem 'Orphic Primal Words':

> As stood the sun to the salute of planets
> Upon the day that gave you to the earth,
> You grew forthwith, and prospered, in your growing
> Heeded the law presiding at your birth.
> Sibyls and prophets told it: You must be
> None but yourself, from self you cannot flee.
> No time there is, no power can decompose
> The minted form that lives and living grows.

For the realisation of his vision, Ibrahim Abouleish did not reckon with years, but with generations and centuries: he envisaged a 200-year plan.

'The minted form that lives and living grows' became a concrete part of the Sekem vision and is one that we work on every day. Continuous development and renewal are essential, especially for community building and social interaction. We believe that a community of people, supported by spiritual ideals, mutual trust and a constant willingness to develop, can create a strong social vitality. This way of life requires a willingness to constantly rethink our points of view – a decidedly uncomfortable endeavour. And this understanding of lifelong learning has allowed Sekem co-workers to continuously integrate intellectual and artistic activities into their everyday work through the so-called Core Program.

The foundations of this development work come from a variety of sources. Of particular importance is Spiral Dynamics, created by Don Beck and Christopher Cowan, which provides a model for the development of individuals, organisations and societies, and the anthroposophy of Rudolf Steiner, which is concerned with the spiritual evolution of the world and humanity. In his many lectures on the spiritual nature of human beings, Steiner described the developmental stages of sentient soul, mind soul and consciousness soul (see the more detailed description in Chapter 9: Our Vision for Egypt 2057 in the Context of Current Events). According to Steiner, all human beings unite these three main states of soul development within themselves, but the qualities that relate to these individual stages can also characterise whole societies and cultures, even different periods of time.

In the Quran, too, we find references to different developmental states of the soul or consciousness, as in the different stages of practising religion or understanding faith. 'Muslim' means the one who professes the religion of Islam, one who is devoted to God and who follows the five pillars of Islam: the creed or declaration of faith, the prescribed prayers, the pilgrimage, the Ramadan fast, and the obligatory social duty. Then there is the 'Mu'min', the convinced believer for whom the search for spiritual knowledge has become an inner concern. And finally, there is the 'Muḥsin', the person who possesses knowledge and judgment, and for whom faith and wisdom have become action. These three distinctions of godliness are referred to in several places in the Quran, for example in Surah 5:93:

> Those who believe and do good works commit no offence
> with regard to what they receive. If only they fear God [know
> with the heart: Muslim] and believe and do good works, and
> further fear God [know in deeds: Mu'min] and believe, and
> then again fear God and are righteous. And Allah loves those
> who do good [act out of love: Muḥsin].

Providing the necessary space for the development of these different stages of consciousness is a supporting pillar of the Sekem vision. It serves its main concern, which is focused on personal growth and the unfolding of individual potential. One of the greatest challenges is to always keep in mind this understanding of the different stages of development, to understand that every single person is at some point on the same path and that everything is in constant change. At each level of consciousness new things are learned and old things are left behind. Only with this realisation is it possible for us to approach another without judgement, but instead learn by observing – and by observing ourselves first and foremost. For the Sekem vision, it is less important how and where development takes place – whether on a spiritual or a physical level – and more about the potential that each human being carries within themselves and which can be realised and put to use.

The four spheres of life

The particular approach taken by Sekem is rooted in the anthroposophical model of the threefold social order. Instead of a centralised society, this model pictures the three spheres of culture (the spiritual), rights (law), and economics – equal in rank but different in essence – as being directed by individuals who are fully active within their respective domain. Each sphere embodies an ideal made famous during the French Revolution: freedom in the spiritual/cultural life, equality in the life of rights, and fraternity in the economic life.

Ibrahim Abouleish embedded these three spheres in the great, all-encompassing sphere of the environment. This created a fourfold division of life, in which the individual spheres are interrelated, individual forms are shaped and enclosed, and the whole can be seen as one great organism. It is of fundamental importance for the vitality and success of the Sekem initiative that these four spheres work in balance with each other, with no particular one dominating the others. The various parts must always be considered in relation to the whole in living interplay.

While this ideal is easy to describe, it is not so easy to put into practice. The demands of the different spheres sometimes conflict. What is considered necessary in business life may not be justifiable in cultural practice. For example, our company NatureTex makes a range of dolls for children. One of the questions we often ask ourselves from a product-development perspective is, 'Which dolls are really beneficial for child development?', as opposed to, 'Which dolls will make the most money?' Another work-related question might concern whether an urgent production issue really is more important than co-workers keeping their Core Program hours.

Similar questions also arise in conflicts within social life. For example, how does a good salary reconcile with the company's economic figures? It is only when those involved are aware that these spheres work together to form a perfect whole that such questions can be answered in a sustainable manner. What is important here is not simply learning to tolerate the demands of the other spheres in each particular case, but consciously perceiving them and giving them our honest attention.

We also find forms of tripartism in Islam. The Islamic concepts of *'alm* (pursuit of knowledge), *'amal* (work), and *mu'amla* (treatment or quality of social interaction) can be related to the different spheres of the threefold social order. *'Alm* is the sphere of culture and education, the pursuit of the spirit. *'Amal* relates to a fair economy and work done in service to the earth and humanity. And *mu'amla* is the social sphere that brings us all together. The connections between these three dimensions in the Quran can be recognised by the fact that all three expressions come from the same root word, which in Arabic indicates a togetherness or an interrelated framework of meaning.

3.

The Sekem Symphony: From Initial Vision to the Present

A vision emerges

In his autobiography, *Sekem*, Ibrahim Abouleish describes how the founding of the Sekem initiative in 1977 – exactly halfway through his life as it would turn out – was of formative importance. He recounts growing up in Egypt, first in the countryside and later in the capital, Cairo. As the son of a successful businessman, he enjoyed a good education and was a particularly curious child: his youthful years were marked by an enthusiasm for art, music, theatre and literature. He developed a keen interest in the works of Goethe, which aroused in him a fascination for European culture and solidified his desire to study in Europe. Against his parents' wishes, he set out at the age of nineteen for the continent that was still unknown to him. In a letter addressed to his father at the time, he described the first beginnings of the vision that would eventually grow into Sekem. He subsequently forgot about this letter until his father returned it to him twenty-five years later, when that vision had begun to materialise. In the letter he had written:

> When I get back ... I will build factories where the people can work, different work than they are used to from farming.

I will build workshops for women and girls, where they
can make clothes and carpets and household goods and
everything else that the people need ... I will establish shops
that sell everything the people need ... I will build a hospital
which I will fill with specialists. In the village I will make a
small quarter for the doctors and their assistants and teachers
to live. I will need teachers as I want to build schools for the
children, from kindergarten to high school.[1]

Ibrahim Abouleish went on to study technical chemistry and
pharmacology in Graz, where he graduated with a doctorate, and it
was there that he met Gudrun Erdinger. They married in 1960 and
had two children: Helmy, who was born in 1961, and Mona, born in
1963. He pursued a successful professional career for many years until
he became acquainted with anthroposophy through the pianist Martha
Werth. This set in motion an inner transformation. With the help of
anthroposophy he worked intensively to understand the more spiritual

Ibrahim Abouleish (centre) with Sekem's first employees in the early 1980s.

forms of Christianity. This in turn led him to a deeper understanding of Islam and a kind of reconciliation of these two great religious streams.

In 1975, Ibrahim Abouleish took his family on a trip to Egypt, where he saw his homeland with new eyes. He was shocked to discover how Egypt, which had still been prosperous in the 1950s, had become a nation full of problems. The population had grown rapidly and the construction of the Aswan Dam meant that agriculture was no longer able to feed the population, despite the heavy use of chemical fertilisers and pesticides. The state's money went to finance wars instead of benefiting the education system.

In the years that followed, a strong desire to return to Egypt matured in Ibrahim. He wanted to share the experiences that had shaped and advanced him in Europe over the past twenty years with the people in his home country. Looking at Egypt's challenges, strongly influenced by his diverse experiences of Eastern and Western cultures and inspired by their spiritual traditions, a vision matured in him:

> I developed a vision of a holistic project able to bring about
> a cultural renewal. As well as the farm it would need one or
> several economic projects, a school and different educational
> institutions and offer cultural projects and medical care.
> My first priority was to educate people. But I would need to
> create concrete institutions for all this so that the project did
> not remain solely an ideal. So I started to look for people to
> work with. I knew I wanted to implement an independent
> project without the help of state funds.[2]

In Europe as well as in Egypt, almost everyone declared his idea crazy. He had turned his back on Europe (and thus on a secure life) to start all over again in Egypt, in the middle of the desert without any security.

Good training, fair working conditions and room for personal development belong to the Sekem economic model.

The vision becomes reality

Ibrahim Abouleish used the biodynamic method of agriculture to revitalise the soil and keep it fertile in a sustainable way. The vision took shape and was given the name Sekem, which referred to the ancient Egyptian term for the life-giving force of the sun. Among the first to join him were Egyptian farmers and field workers, to whom he conveyed the meaning of biodynamic agriculture with the help of Islam. Medicinal herbs were cultivated to make teas and herbal medicines, and companies sprang up to process the field produce into healthy, high-quality food and other products. Soon, the hustle and bustle of the new Sekem farm became even more colourful. The workers' children and those from the surrounding villages attended the schools and training workshops that Ibrahim gradually established with the help of his friends. More people joined the project – from Egypt but also from Europe – and that was a blessing, because the vision needed people to become a reality.

Today, the desert is thriving. Many hundreds of people now live, learn and work on around 200 hectares (495 acres) of formerly barren land near the town of Bilbeis, some 70 kilometres (45 miles) north-east of Cairo.

With people came growth, successes and failures. One of many examples of Sekem's success is the cotton story. In the late 1980s, more and more pesticide residues were found in Sekem herbs, which were grown under strict biodynamic guidelines. It became clear that these came from aeroplanes spraying pesticides over nearby cotton fields to control pests. We were initially unable to convince the Egyptian Ministry of Agriculture how harmful these chemicals are to the environment and to people. The only way to put an end to this practice was to grow organic cotton ourselves and prove that this could be done successfully without the use of pesticides. Together with international and national experts, research was conducted for several years in the laboratory and later in the field until Sekem was able to prove that unsprayed organic cotton could produce as much revenue as conventional cotton. As a result, the Egyptian Ministry of Agriculture was persuaded to ban the spraying of chemical pesticides from aeroplanes, reducing the annual use of pesticides in Egypt by over 90%. The large quantities of high-quality organic cotton that Sekem had grown were self-processed instead of being sent for conventional processing, which would have involved the use of toxic substances such as chemical dyes. This led to the foundation of the Sekem company NatureTex as well as the Egyptian Biodynamic Association (EBDA), which to this day supports farmers throughout the country in converting from conventional to biodynamic agriculture.

This success was also the first clear sign that Ibrahim Abouleish's seemingly utopian plans could become a reality and work in practice. In all spheres, in the economy as well as in culture, Sekem's successful developments are an example of how important a network is. People all over the world support Sekem's cause in different ways. For decades, many activities have been supported by the Sekem associations in Europe with great commitment and out of a deep solidarity with our ideals. Many of our business partners stand by our side in true partnership. It became clear just how important these reliable business partners are during and after the revolution in Egypt in 2011.

In 2012, despite the great political crisis occurring at the time, the Heliopolis University for Sustainable Development, initiated by Sekem, finally opened its first three faculties. Ibrahim Abouleish had always been keen to develop a university that would combine teaching and applied research for the future of young people. He once wrote in a newsletter:

> Sekem must reach a critical size so that this idea can be
> grasped. This includes a university because there people
> can learn to do research, they can come to terms with the
> Sekem idea. You can then show scientifically that biodynamic
> agriculture makes the soil fertile.

The many international honours that Sekem has received since the turn of the millennium has contributed to the grasping of this idea. In particular, the award of the alternative Nobel Prize (the Right Livelihood Award) in 2003 supported the spread of the Sekem vision throughout the world.

Year of destiny: 2017

A fateful year for Sekem came in 2017. First, there were two big celebrations in the spring: the fortieth anniversary of the founding of the Sekem initiative and Ibrahim Abouleish's eightieth birthday. Friends came from all over the world to offer their congratulations to the founder, whose life's work not only transformed the desert and made it flourish, but also shaped the lives of many people in various fields.

Festival celebrations have been cultivated in Sekem from the very beginning as a way for us to come together in appreciation of what has been achieved and to anticipate what is to come. All of these celebrations have been beautiful, but the two in 2017 were particularly outstanding because they were anniversaries. They were planned thoroughly and the fact that they were both significant milestones made them seem joined together by an invisible hand. They were worthy celebrations in the highest sense.

A few months later, on June 15, 2017, Ibrahim Abouleish passed away.

That year was also fateful because we, the Sekem community and especially the Sekem Future Council, saw the death of Ibrahim Abouleish as a call to continue his life's work. Within days of his funeral we were already discussing the way forward.

We realised that we wanted to develop an expanded vision for the next forty years, one that already had a firm foundation in what had been developed during the first forty years. With this task in hand, we occupied ourselves for a year so that, on the first anniversary of Ibrahim Abouleish's death in June 2018, we could publish our vision of Egypt in 2057.

Dr Ibrahim Abouleish died in 2017 at the age of 80.

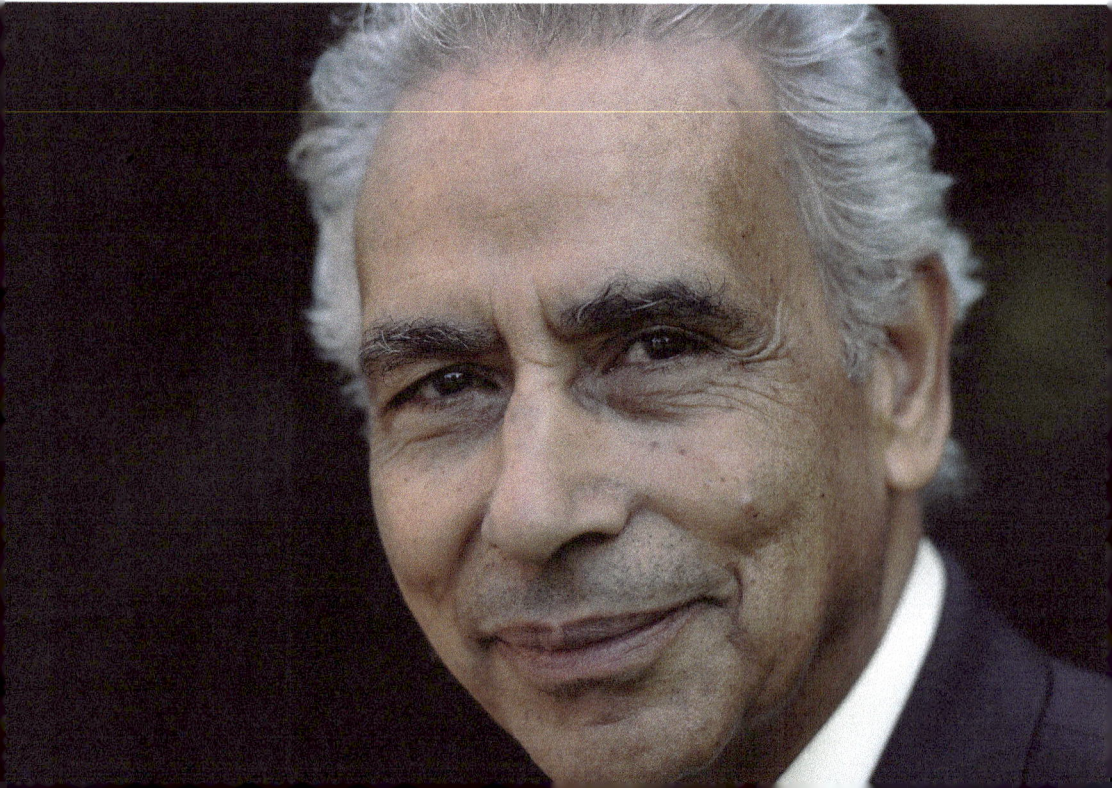

A vision for the future

As soon as the spirit is directed towards a goal, many things come to meet it.

Johann Wolfgang von Goethe

With the death of Ibrahim Abouleish, it became clear that the role of the founder and visionary could not be replaced. But his vision can be furthered by those who lead Sekem into the future. This requires the firm foundation of the original vision and its spiritual sources, but also the ability to adapt it to the needs of people today. So we drew inspiration from the original vision and, like Ibrahim Abouleish, took a close look at the current challenges. We tried to see what we could carry forward and what would come to meet us from the future. We realised that our task now is to make available on a larger scale what has been achieved and tested in Sekem. In doing so we are not interested in further growth in the sense of more and bigger, but in making the substance of our work so broadly applicable by means of concrete models that real social and systemic change can come about.

Thus, we have formulated a vision that describes our image of Egypt in another forty years, in 2057. We intend to use the first ten years, until 2027, to realise the vision goals for Egypt within Sekem, because here in our own hands we have everything necessary for their implementation.

The starting point for our vision of Egypt in 2057 may sound as absurd today as the vision of a sustainable initiative in the desert once did when Sekem was founded. In the meantime, however, we know that visions of this kind carry incredible potential. And we are aware of how existential our fundamental trust is. In the beginning, there was no master plan, just an idea that perhaps could be put into action if the circumstances were in any way right. The idea might fail or it might succeed, but as long as the requisite work was done, the way forward would be found.

Part 2:
The Sekem Vision Goals

4.

An Overview of the Sekem Vision Goals

In order to realise our vision for Egypt in 2057 and our intermediate vision for Sekem in 2027, we have formulated goals in the following four spheres: culture, ecology, the economy and social life. Since we first published these goals in 2018, they have undergone several revisions and they will continue to be adapted in the coming years. What follows is a brief overview of our Sekem Vision Goals. They will be explained in more detail throughout the book.

Culture

- **Vision Goal 1: Lifelong learning** – individual development is the central concern of Egypt's educational systems. The developmental potential of each person is promoted holistically and throughout life.
- **Vision Goal 2: Holistic research** – a holistic research model is developed that incorporates natural science, the humanities and spiritual science.
- **Vision Goal 3: Integrative health** – integrative health and therapy practices are established and widespread throughout the Egyptian healthcare sector.
- **Vision Goal 4: Arts and culture** – regional, local and international arts and cultural activities are valued by and co-created with the Egyptian people.

Ecology

- **Vision Goal 5: Agriculture** – Egyptian farmers are implementing sustainable agriculture using biodynamic or organic methods.
- **Vision Goal 6: Sustainable water management** – Egypt operates a sustainable water-management system that minimises water consumption, reuses wastewater, and uses innovative water-harvesting and irrigation techniques.
- **Vision Goal 7: Renewable energy** – sustainable energy management based on renewable energy and optimised consumption is being implemented in Egypt.
- **Vision Goal 8: Biodiversity** – Egypt's biodiversity is sustainable, growing and flourishing.
- **Vision Goal 9: Climate neutrality** – Egypt is climate neutral because no CO_2 is emitted that nature cannot reabsorb.

Economy

- **Vision Goal 10: Circular economy** – businesses in the country practise a circular economy. Egypt is a showcase for waste-reduction management.
- **Vision Goal 11: The economy of love** – businesses in Egypt operate according to the principles of the 'economy of love', emphasising transparency and considering true costs..
- **Vision Goal 12: Ethical banking and finance** – Egypt has implemented ethical banking and finance.
- **Vision Goal 13: Sustainable lifestyle** – responsible consumption and a sustainable lifestyle are mainstream. There is a wide range of sustainable products and services for all customer needs and social classes.

Social life

- **Vision Goal 14: Agile organisational structures** – across various organisations in Egypt there are vibrant and agile organisational management practices that are tailored to people's development and state of consciousness.

- **Vision Goal 15: Diversity and equal opportunities** – Egypt celebrates diversity and ensures equal opportunities for all, regardless of age, nationality, religion or gender.
- **Vision Goal 16: Social transformation** – social transformation in Egypt has led to sustainable rural development and regenerative cities, empowering people to take responsibility for shaping the country's future.

Within each sphere we have established centres of applied research with experts who are entrusted with the realisation of their particular goals. In these centres prototypes of models and tools are developed and tested. These models are created in such a way that they are easy to adopt, giving them system-wide relevance.

The Sekem Vision Goals will be implemented over five phases:

1. **Vision finding**: this is carried out in the various Vision Groups and through the work of the Sekem Future Council, as well as on individual bases.
2. **Research**: this is carried out at Heliopolis University, in Vision Groups at Sekem, and with partner organisations.
3. **Prototyping**: this is carried out at Heliopolis University and Sekem, and with partner organisations.
4. **Upscaling**: this is carried out in the thirteen villages surrounding Sekem and at Sekem Wahat Farm.
5. **Mainstreaming**: the solutions and practical models we develop are ultimately established throughout Egypt.

With some goals, we are already at the point where they are driving social, ecological or economic change. This applies, for example, to the spread of organic agriculture, mainstream models for the use of renewable energy or changes in the education system. Other goals are still in the research or model-development phases.

The 'Vision Circle' presents the sixteen Sekem Vision Goals with each segment representing one goal. The more an individual segment is filled with colour, the closer that goal is to being achieved. In the following chapters, each goal will be discussed in more detail.

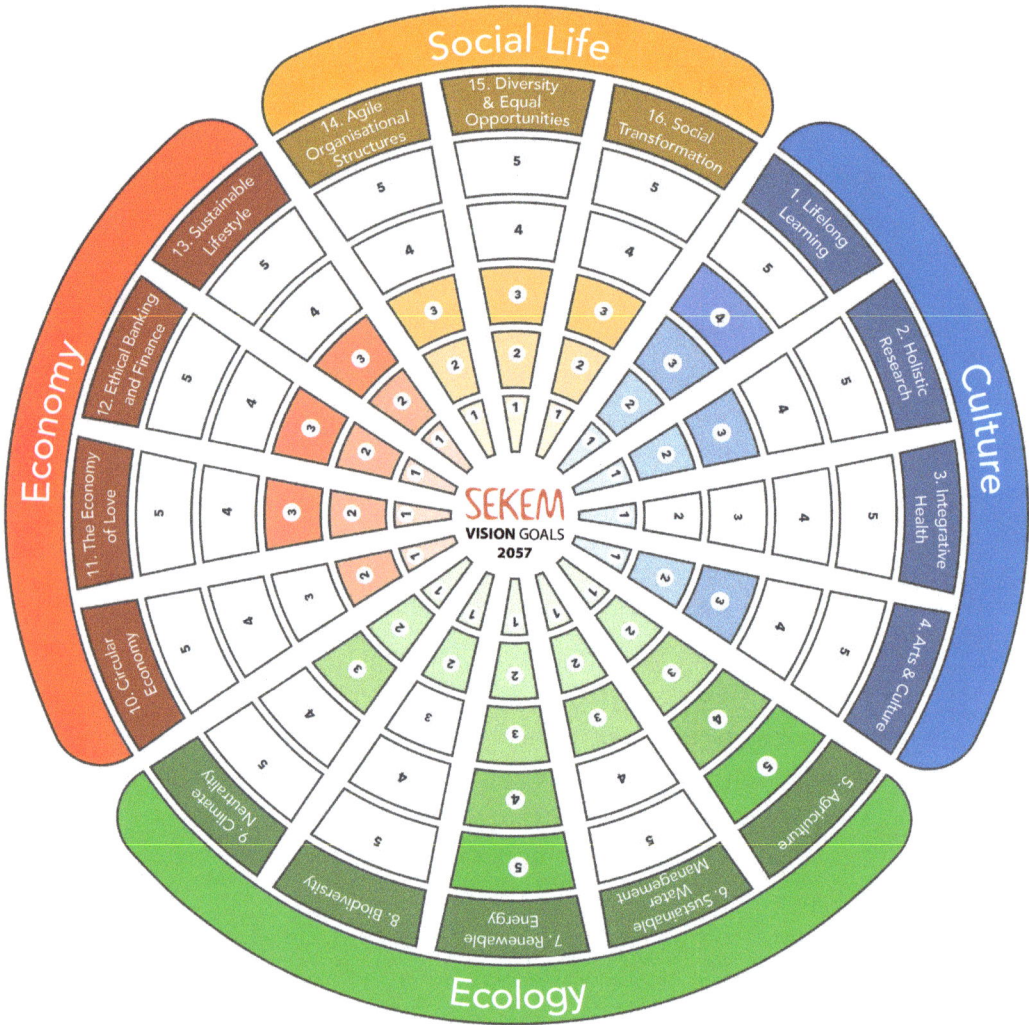

The 'Vision Circle' provides an overview of the progress of the sixteen Sekem Vision Goals towards realisation.

5.

Culture

It has not rained light for days
The wells in many eyes
Are tormented by drought.
That is why friends
Are not easy to find
In this wasteland.
Where almost everyone has fallen ill
From jealously watching
Nothingness.
On this caravan
Through blistering desert heat
Careers and cities can seem real.
But I say to those close to me:
'Don't get lost in them,
There it has not rained light for days.
Look, almost everyone is sick
Of loving
Nothingness.'

Shams Ed-din Al-Hafiz

Inspirations for lifelong learning

For the Sekem vision, cultural life includes all areas that contribute to individual development: from education, art and aesthetics to topics such health and religion. In many approaches to sustainable development, the cultural dimension is neglected or not considered at all. In Sekem, it has the same importance as the spheres relating to

In Sekem's curative education facility, people with special educational needs can develop their potential.

ecology, the economy and social life. For Ibrahim Abouleish, this was fundamental since it is through the cultural sphere that individual development is nourished spiritually and emotionally; only in this way can each human being grasp and develop their own wholeness. Ibrahim Abouleish's first two investments when he established Sekem are typical of this attitude: he bought a tractor and a grand piano. As he wrote in his autobiography:

> The artistic appearance of Sekem was of prime importance to me. I wanted beauty and grace not just in addition to the companies, but as an integral part from the start spreading its influence over everything.[1]

For Ibrahim Abouleish, the great importance of the cultural dimension was rooted in the conviction that without self-knowledge, which must include the emotional and spiritual life, social action and development are not possible. The spirit must be nourished, but this cannot be done through education and science alone. The experience of art and beauty is equally important. Ibrahim Abouleish was convinced that the world reveals itself to people through the arts. He himself resumed violin lessons at the age of seventy-five.

At Sekem, we promote art and culture in many different ways: from the aesthetic design of buildings and spaces, to the creative courses in which all members of the Sekem community regularly participate, to cultural events and festivals.

Rudolf Steiner wrote, 'The process of knowledge is the process of development towards freedom'.[2] Through this freedom, people are able to find the meaning of their lives and thus their role in the world. In the Quran, too, the importance of gaining knowledge is emphasised many times. For example, Surah 94 states:

> Have we not widened your breast, and freed you from your burden, a burden that bent your back? Did we not increase your capacity for knowledge? Verily, in the heavy lies the light. Yes, in the heavy lies the light. When you have realised this, then act! And aspire to your Lord.

This is the same path of development mentioned by Steiner.

Theatre performance by and for members of the Sekem community.

The audience at one of Sekem's many cultural performances.

A constant striving for wisdom

Ibrahim Abouleish often referred to a task he saw for Sekem, and to which his religion called him, which was to encourage lifelong learning in people. Again and again, the Quran calls for lifelong learning, such as in the words, 'My Lord, increase my knowledge' (Surah 20:114). These statements not only serve the teachers of the Sekem school as guiding principles for an education in freedom, but also accompany the entire community in their daily activities.

The Quran calls on us to always strive for wisdom. Wisdom is a gift of God, which the believer asks from God. As one Hadith says, 'All wisdom is the lost property of the believer, wherever he finds it, he should take it up.'[3] This means that the acquisition of knowledge is not limited to a particular environment, such as a school, or to certain formats or teachers. It can take place anywhere and throughout one's life. At the same time, however, it requires effort on our part, as stated in the aforementioned caveat of Surah 13: 'Allah does not change the condition of people until they themselves change their condition' (Surah 13:11).

We are thus called to set out on the path of learning, as expressed in another Hadith:

> I heard the Messenger of Allah say: 'If one goes on a path of
> seeking knowledge, Allah will make him go on one of the
> paths in Paradise. The angels will lower their wings in great
> joy over the seeker of knowledge, and the dwellers of the
> heavens and the earth and the fish in the depths of the water
> will beg forgiveness for the scholar. The knower stands above
> the [common] believers like the full moon at night above
> the rest of the stars. The knowledgeable are the heirs of the
> prophets, and the prophets leave neither dinars nor dirhams:
> they leave only knowledge, and whoever absorbs it will have
> abundance.'[4]

Lifelong learning is deeply rooted in the Sekem vision and in Sekem's daily life. It begins with the school facilities, which we design so that children can become free-thinking and free-acting adults. This remains a particular challenge because current educational systems neglect to foster important skills of the heart and hands. It is primarily the head, the intellect of children, that is addressed. The need to involve not only the head in learning but also the heart is found in the Quran:

> And Allah brought you forth from the wombs of your
> mothers while you were ignorant. And He has made you
> hearing, eyes and hearts, that you may be grateful.
> (Surah 16:78)

We want education to nurture the essential nature of learners and allow their creative talents to develop freely. Thus, Sekem schools do not exclusively follow the state curriculum, which in Egypt is mainly focused on memorising information. Our teachers are trained to use a variety of educational elements to convey experiences, to enable individual interactions with the subject matter and, in particular, to evoke questions in the students. Wherever possible, we integrate artistic and craft subjects with practical activities.

Surah 22:46 also gives information about the importance of the powers of the heart in relation to the acquisition of knowledge and cognition:

> Have they not then gone about the earth, that they may grow hearts with which to understand, or ears with which to hear? For the eyes do not go blind, but the hearts that are in the breast go blind.

If we learn only intellectual concepts, we connect with the world only in a purely rational way. This leads to a cool, more distant connection, which accordingly influences the life of society. Through the involvement of the various arts, the ability to perceive and feel is trained and practised. Ultimately, one can only really connect with something and develop a deeper understanding through the powers of the heart. Neglecting the life of feeling has instead exacerbated many of the problems facing the world today. Consider climate change, for example. Many people nowadays understand with their intellect that climate change is the greatest threat to humanity, and yet, even though we are reminded of this every day, very few of us act accordingly. We have not developed a deep enough emotional understanding of the issue because we view the challenge almost exclusively in abstract scientific terms.

At Sekem, we assume that the interplay of artistic, emotional and scientific dimensions will play an important role in the future of learning and research.

The Sekem schools are inspired by Waldorf education, which is based on the threefold division of the human being into spirit, soul and body, and the associated faculties of thinking, feeling and willing. The seven-year rhythms in the human life cycle described by Rudolf Steiner are taken into account in child development and in pedagogy.

The number seven also plays a role in the Quran, especially when it comes to the seven heavens and the seven earths. In Surah 71:13–15, for example, seven layers are specifically addressed in connection with stages of human development:

What is the matter with you that you do not worship
Allah in reverence, when He has created you in stages of
development? Have you not seen how Allah created seven
heavens in layers?

It is against this background and with similar references from the
Quran (Surah 17:44; Surah 65:12; Surah 67:3), that we view the seven-
year rhythms in the development of children in pedagogy.

The pursuit of self-knowledge

At Sekem, the pursuit of knowledge and lifelong learning is not limited
to the pupils and students of the Sekem schools or those involved
in scientific research at the Heliopolis University for Sustainable
Development. It includes all community members. As we have already
seen, in the first Islamic revelation Allah asks Muhammad to acquire
knowledge by telling him to read.[5] In addition to numerous other
references concerning the importance of knowledge, the Quran also

Students at Heliopolis University for Sustainable Development.

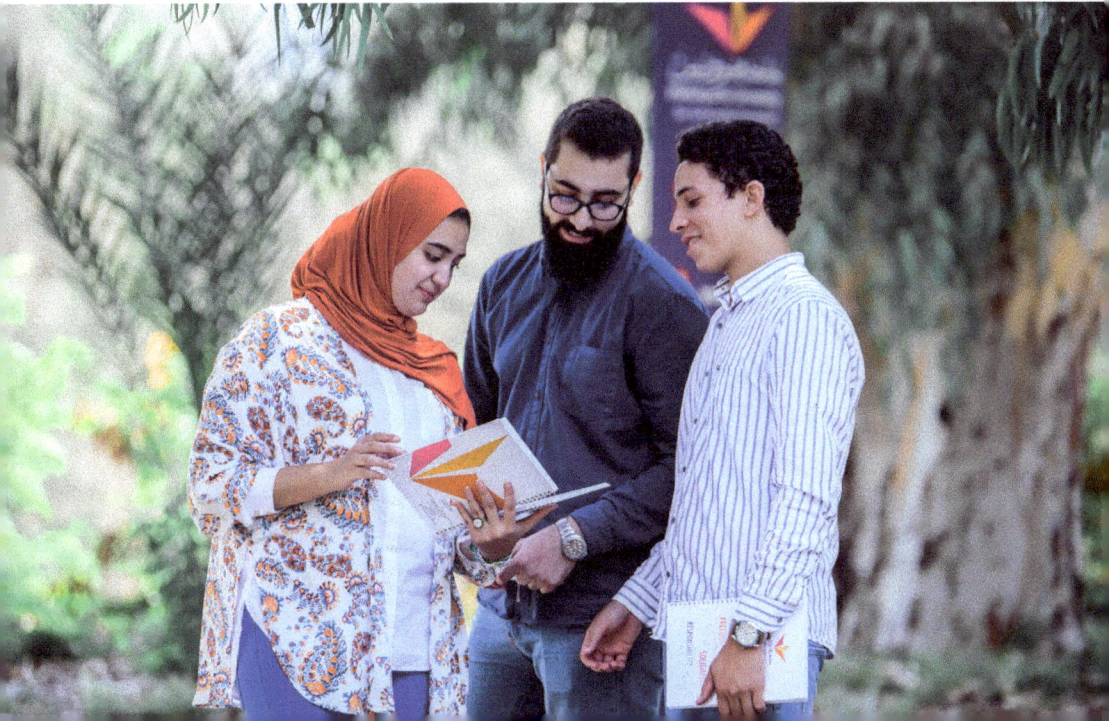

appeals to human beings' use of reason when it says in Surah 7:184:
'Do they not think about it?' And there is an explicit reference to the
importance of lifelong learning in Surah 35:28: 'Reverence for Allah
certainly has among its servants the scientists.'

Ibrahim Abouleish was himself a scientist and researcher with a
doctorate, and he was keen to apply and verify all theoretical knowledge
in practice. He was enthusiastic about how Goethe combined science
and art, and so it was a matter of course in Sekem from the very
beginning that research into the various spheres of life, into solutions
to current questions and future challenges, was always included. In this
regard, Sekem's comprehensive structure offers the opportunity to use
theoretical research approaches taken from everyday life and test them
in practice.

It was Ibrahim Abouleish's belief that education should not only
shape the individual but that the individual should themselves become
a source of education. It was not enough for them to be cultured and
educated, they should want develop culture further.

In this context, the ninety-nine names of Allah played a significant
role for Ibrahim Abouleish. They served as a model for him. To live
up to these names, each of them attributes of God and representing an
ideal, became for him a part of his own striving for development and
knowledge. Just as these ideals were of great importance in his life, they
are also deeply woven into the activities of the Sekem initiatives. One
of the names of God is 'The Light'. Surah 24:35 is also called the 'light
verse' and is still considered one of the most enigmatic Quranic verses
in Islamic studies:

> Allah is the Light of the heavens and the earth; His light is
> comparable to a vessel in which there is a lamp. The lamp is
> in a glass. The glass is as if it were a twinkling star. It is fed
> by a blessed tree, an olive tree, neither eastern nor western,
> whose oil almost glows even without fire touching it. Light
> upon light! Allah guides to His light the one who wills, and
> Allah guides to man the parables. And Allah knows about all
> things.

The parables and images of this verse are considered to be the roots of Islamic mysticism, and Ibrahim Abouleish saw in it a clear connection to the anthroposophical view of the human being. In the verse, the light of Allah is compared to a vessel containing a lamp. In this vessel, Ibrahim Abouleish saw a reference to the physical body. The lamp itself is enclosed in glass and the glass appears as a twinkling star. In the sparkling light he saw a symbol of the soul and the sensations that flicker within it. The lamp receives the divine light from an olive tree, which is 'neither eastern nor western' – that is, not earthly. The blessed tree, symbolising the life forces, represents the etheric or life body, the spiritual body possessed by all living beings that is the bearer of life. Finally, the individual 'I' is the inner light, which resembles the oil that almost glows without being touched by the fire. Thus, Ibrahim Abouleish recognised in this verse of light an image of the human being in which the possession of knowledge (the inner light) is emphasised. In Sekem we understand this call to possess knowledge as a call to know ourselves.

Beauty and the spiritual life

Ibrahim Abouleish's main concern was to give a space to beauty in Sekem. He was convinced that in beauty, the spiritual life and the life of feeling reveal themselves, thus creating a basis for development. As Friedrich Schiller said in his poem 'The Artists': 'It was only through beauty's morning gate that you gained the land of knowledge.'[6]

Through beauty and art, a path to the divine can be found. In Islamic architecture and calligraphy we find numerous references to this, for example in the Hadith: 'God is beautiful and loves everything beautiful.'[7] Calligraphy is an art form that attempts to make linguistic beauty, the sound of God's words, visible. The Quran itself is a work of art because of its poetic text and the possibility of recitation. Surah 39:23 points out:

Allah has brought down the most beautiful tidings, a book of similar, repetitive verses, from which the skin of those who fear their Lord shudders. Then their skin and their hearts soften and incline to the remembrance of Allah.

In this surah, different qualities of beauty are expressed. The 'most beautiful tidings' refer to the good or beautiful content of something; 'similar, repetitive verses' refers to the beauty of something that is balanced and in harmony, and then there is the beauty that makes 'hearts soften'. At Sekem, we strive to include these three types of beauty in our everyday activities: doing good deeds for people (love); learning to create balance, such as with nature, and the use of art and culture to touch the heart, meaning to go beyond the material and the intellect. Thus, one of the many descriptions of Sekem that Ibrahim Abouleish gave was:

> I see Sekem as a beautiful painting and the farm as the suitable frame. This painting is enlivened by the different colour nuances which the Sekem employees bring to it ... Through our artistic deeds a beautiful garden is created.[8]

As he also explained:

> The power of art gives hope and courage and through beauty works in a humanising fashion ... Art also leads to a sensory training, through which people advance themselves and liberate their senses.[9]

One of many examples of how we put this into practice is through the art of eurythmy, a form of movement created by Rudolf Steiner to express language and music. This is an integral part of the curriculum in the Sekem schools and at Heliopolis University, as well as in staff training. There are also regular public eurythmy performances. Ibrahim Abouleish had already identified the forms and movements of eurythmy on reliefs in ancient Egypt and thus defined it as an important aesthetic element for Sekem.

Eurythmy can also be associated with the Quran, through the so-called 'mysterious letters'. Twenty-nine Quranic surahs begin with the mention of one or more unconnected letters. So far, Islamic scholars have not been able to discern the meaning of these letters. We have taken a closer look at them and their sequences with the help of eurythmy.

The eurythmic gesture of the letters can lead to a deeper understanding of the essence of a letter and its expressiveness, and thus also give information about the connection to the content of the surah. As an example, Surah 68, *Al-Qalam*, begins with the letter 'Nun' (N) and speaks of the processes of cognition and writing down spiritual content. The eurythmic movement of 'N' involves turning towards an object, touching it and then detaching and withdrawing from it again. This also reflects what is happening in the surah: condensing a spiritual process, writing it down and detaching from it again. This becomes a process of cognition: we have to connect with something and then distance ourselves from it again in order to come to understanding.

Hygiene

Hygiene, health and healing are also addressed many times in the Quran as important aspects of well-being and development. There is a general obligation to hygiene and cleanliness, for example, with specific details on regular ablutions. Before each of the five daily prayers, believers are meant to wash extensively with pure water:

> O you who believe, when you stand for prayer, wash your
> faces and your hands beforehand up to the elbows and wipe
> your heads, and wash your feet up to the ankles. And if you
> are sexually defiled, purify yourselves. And when you are sick
> or traveling, or when one of you comes from the privy, or
> when you have touched the women and you find no water,
> then seek a clean ground and wipe your faces and hands
> therefrom. Allah does not want to impose affliction on you,
> but He wants to make you clean and complete His mercy on
> you, that you may be grateful. (Surah 5:6)

This hygienic advice is important for the prevention of diseases, but health is also addressed as a holistic concept. The Quran itself is described as a means of healing and promoting health: 'Down we send from the Quran what is healing and grace for the believers' (Surah 17:82). In it we see indications that health cannot be defined solely by the absence of disease and that more than physical criteria contribute to the maintenance of health, such as peace, meaningfulness and a just social environment – these are the good messages of the Quran. An example of this is also given in the Arabic expression for recovery. When Muslims wish for someone to get well, they wish them peace: *salamtak* or *alf salama*, from the Arabic *salam*, used in greeting, which literally means 'peace'.

Ibrahim Abouleish was concerned with health as a Muslim, a pharmacist and a chemist. He knew that, according to the Quran, humanity was created in an ideal form.[10] Thus, he understood health in the Islamic sense as a gift from God, which should be preserved and cared for. Health and wholeness, as well the effective treatment of diseases, occur when the human body is recognised and treated not as an automated system, but as a divine miracle consisting of body, soul and spirit, endowed with thinking, feeling and willing. This is also how Rudolf Steiner viewed it:

> Human beings are what they are through physical body, ether body, soul (astral body) and 'I' (spirit). In health human beings must be considered in terms of these aspects, in sickness perceived in terms of balance between them being upset; for health, it is necessary to find medicines that will restore the upset balance.[11]

On this basis, we in Sekem have resolved to focus on the preservation of holistically understood health, so that illnesses are less likely to arise and spread in the first place.

Forty years of culture in the desert

Situated among the fields and buildings of the Sekem Mother Farm is a large amphitheatre fashioned from hardened local clay. Farmers, co-workers, students and friends gather here from all over the world at least twice a year. Members of the Sekem community design and stage cultural programmes themselves. Co-workers present their daily work

The amphitheatre on the Sekem Mother Farm is a popular venue for cultural events and festivals.

in a skit, school children sing or the school orchestra performs musical pieces. Cultural activities have become a natural part of all Sekem institutions.

We have been able to establish a variety of educational institutions: baby groups and kindergartens, schools for general education, a vocational training centre and a curative education facility. An outpatient clinic for the population of the surrounding villages is also located under the umbrella of the Sekem initiative. In these facilities, people are given the opportunity to develop in a healthy and free manner, and to realise their full potential. Emphasis is always placed on holistic methods.

This approach continues in the Core Program in adult education. This programme is a fixed component for all the staff of Sekem-affiliated institutions. It includes a wide variety of artistic and cultural activities as well as regular intellectual exchange on socially relevant topics, joint sports or professional training. Students at Heliopolis University also have a Core Program and cannot finish their studies unless they also successfully complete the programme's designated seminars. Subjects include art, sociology, philosophy, Egyptology and ecology. In this way, a view of the wider world is opened up, a sense of social responsibility is awakened and a desire to help shape society is promoted.

Since 2012, Heliopolis University for Sustainable Development has grown steadily and now operates five faculties: Business, Engineering, Pharmacy, Organic Agriculture and Physiotherapy. The university aims to educate young people from all socio-economic backgrounds to become creative and innovative leaders, empowering them to recognise current and future challenges and to address them holistically. In 2021, the Ministry of Education granted permission for another Heliopolis University campus with eight additional faculties. This new campus will open in a rural setting near the Sekem Mother Farm, giving young people who do not come from metropolitan Cairo the opportunity to benefit from our alternative approaches to higher education. In the future the thirteen faculties will include humanities and social sciences as well as an arts degree programme.

Teaching children music helps them develop more than just intellectual skills.

Challenges and opportunities

Over the second half of the twentieth century, Egypt's cultural life, educational system and healthcare steadily deteriorated. The reasons for this include rapid population growth, political restrictions and a lack of awareness of the importance of a vibrant cultural life for a country's progressive development.

State schools are free in Egypt, but the quality of education is so poor that every family for whom it is financially possible enrols their children in a private school. Classes in public schools are overcrowded. It is not uncommon for one teacher to teach up to seventy students at a time. The teaching staff is underpaid and therefore usually forced to work a second job as a tutor. The curricula for mathematical and technical subjects consists of rote learning, with students made to memorise facts. In addition, the schools and their surroundings are often in terrible condition: facilities are old and falling apart, and hygiene standards are lacking. Despite recent reform attempts in selected provinces, the Egyptian school system has repeatedly been rated as one of the worst in international comparisons. The state is continually trying to improve the situation, but with little success so far because of the enormous challenges. The percentage of illiteracy among women and men over the age of fifteen is estimated at over 27%. However, Egypt has a large population of young people (52% under the age of twenty-five in 2020), in which much hope is placed.

Universities are plentiful, but few have teaching models that prepare students for the challenges of the future. Young people's choices of subjects are based more on prestige than on individual interest or talent. Apprenticeships that focus on crafts or farming enjoy little standing. As a result, these and other trades programmes tend to be under-resourced and lack skilled instructors.

Like the education system, Egypt's healthcare suffers from limited funding and inadequate facilities and medicines. Well-trained specialists tend to emigrate. Although basic state health insurance is guaranteed for all, it covers only a few services. Medical treatments tend to be symptom-oriented, with extensive use of strong medications. For example, antibiotics are available without a prescription and are

Creative activities play an important part in adult training courses.

over-used. The weaknesses of this approach, the widespread lack of exercise and the absence of a balanced diet in Egypt contribute to high incidences of chronic diseases. Diabetes, obesity and hypertension are common.

Until the 1950s, Egypt had a rich cultural life and was known throughout the Arabic-speaking world as a centre for artistic and cultural activities, with artists such as the legendary singer Umm Kulthum. Ibrahim Abouleish liked to tell how in his youth people sat in front of their radios whenever Umm Kulthum sang. Whether rich or poor, and regardless of their level of education, people loved the music of the time. Today, the opera house in Cairo, a city with a population of over 20 million, is rarely sold out, and venues for the country's diverse cultural treasures often give way to amusement parks or digital entertainment. In our neighbouring town of Bilbeis, for example, which has a population of over 400,000, there is only one performance venue, and it is rarely used.

Even the Egyptian winner of the Nobel Prize in Literature, Naguib Mahfouz, who was awarded the prize in 1988 (the first and so far only Arabic-speaking writer to be honoured), was hardly followed by a list of similarly renowned names. The number of books published annually in Egypt is considered very low by international standards. The majority of Egyptians spend their free time in front of the television or on the Internet watching translated programmes from the West that are far from the reality of their own lives.

Egypt does have a relatively large number of researchers and graduates. But the number of publications is low, and few of the innovations mentioned in them are used in daily life. Research is stagnant and hardly integrates new ideas and approaches. To make matters worse, although religion and spirituality are important to most people, they are difficult to reconcile with the natural sciences. As a private person, one is deeply religious, but in the laboratory, non-materialistic perspectives are excluded: no criteria other than the purely scientific are considered.

We believe that without a vibrant cultural life there is no chance for sustainable development. Egypt urgently needs a better education system, modern research and more cultural events.

Vision Goal 1: Lifelong learning

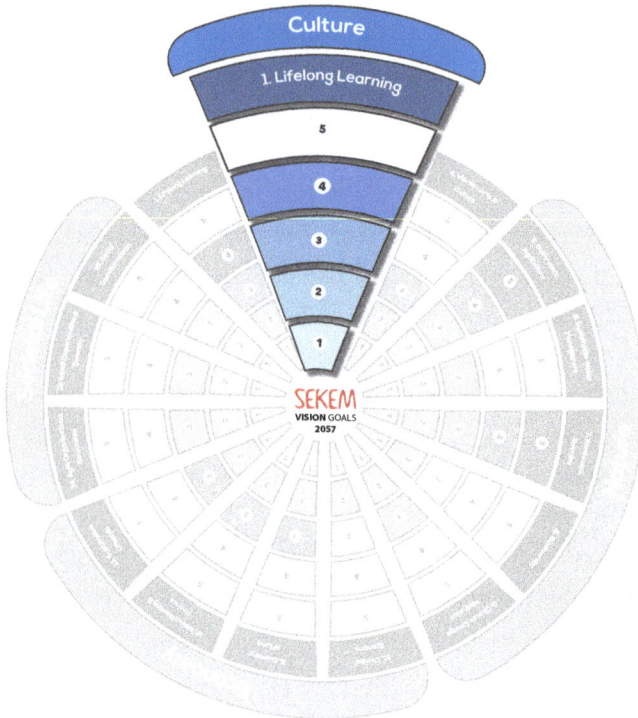

Vision for Egypt 2057

Individual development is the central concern of Egypt's educational system. The developmental potential of each person is promoted holistically and throughout life.

Progress

We are currently in an intensive phase of developing new models and making them relevant for Egyptian society before distributing them. We have 18 ongoing projects and activities, including:

- community-based learning programmes for students
- adult education for sustainable development

Our Vision Circle for the organisation of cultural life focuses on the following question:

- How can we provide cultural support so that people can unfold their own potential instead of just 'training' them?

The results of this work are applied in practice at our Centre of Excellence for Education for Sustainable Development. Here, scientists from a wide range of fields develop alternative and innovative concepts that create the appropriate conditions in educational institutions, work environments and everyday life for people to develop their individual abilities.

Over the past few decades, teachers at the Sekem schools have worked extensively to create spaces for the development of potential. They can now draw on a wealth of experience to help their new colleagues in developing curricula. An important example of this is the balanced approach taken towards the promotion of knowledge, arts and crafts: the intellect, the emotional life and practical skills and activities are addressed in equal measure. We want to make the curricula widely available and design them in such a way that they can be used nationwide in both public and private schools.

For this goal, we are already in the distribution phase. The introduction of a curriculum inspired by the Sekem schools has begun in several schools in the area. Another step in this direction has been taken with the Education for Sustainable Development kits: textbooks and comprehensive school materials published by Sekem in partnership with other institutions that are now available to all schools in the country.

Even in the field of vocational education the curricula are designed to serve other institutions in the country. For example, our curriculum for biodynamic agriculture is approved throughout Egypt.

In the country at large, training and working conditions for instructors rarely focus on supporting the development of student potential. At Sekem, however, we offer training programmes for university lecturers on the need for holistic pedagogy and how to implement it in the university. Professors and lecturers can complete their obligatory three-year pedagogical diploma with us. One of our successful modules, the 'Pedagogical Conference', holds weekly sessions for the teaching staff of Heliopolis University. Here, socially relevant issues as well as cultural topics are discussed in connection with pedagogical approaches. It is our aim to introduce these and other measures that promote individual development to universities across the country.

Likewise, we want to use experiences gained by our staff from the Core Program to design programmes that can be adapted according to the different needs and contexts in which they are used. For example, Core Program activities are designed differently for farmers than for sales representatives or creative professionals. The goal, therefore, is to design Core Program models that can be used in government offices as well as in industry or in the private sector.

Inspiration: learning from the community

All-round, lifelong learning is only possible if we engage with the world in a very tangible way, connecting with our hearts as well as with our minds. As part of our Community Based Learning Module, all students spend several days on our desert farm in Wahat El-Bahareyya and in one of the thirteen villages located around the Sekem Mother Farm. There they perform a kind of community service: they set up systems for waste separation and recycling, organise cultural events, plant trees and experience first-hand what it means to produce food. This is intended to bring learning out of lecture halls and classrooms and into the world in a practical way. Most of the young people return to their studies full of inspiration and enthusiasm.

Vision Goal 2: Holistic research

Vision for Egypt 2057

A holistic research model will be developed that incorporates natural science, the humanities and spiritual science.

Progress

Our research is influenced by the results of natural science. It is also clear to us that the spiritual-scientific view of nature and humanity is just as relevant. We therefore want to work with those involved in spiritual science to develop holistic research methods that we can apply ourselves. We have 4 ongoing projects and activities, including:
- a planned biocrystallisation laboratory

The Heliopolis University conducts research in many areas, including renewable energy, water, soil fertility and food vitality.

In the future, our applied research on herbal medicines or biological seeds will be increasingly holistic. For example, we have started to work on making the vitality of food visible through the process of sensitive crystallisation (see below). We want to extend this approach to other areas of research.

We have also been running a series of trials for several years that compare conventional, organic and biodynamic methods of agriculture. With these trials, we hope to demonstrate to farmers and consumers that organic and biodynamic agriculture has a positive and sustainable impact on soil, biodiversity and yields. We also conduct research on methods of CO_2 sequestration, water use and renewable energy.

We believe that a holistic research method – one that is able to grapple with complex systems and understands each component part in relation to every other part as well as to the whole – will make an important contribution to raising people's awareness and thus shape and support a change towards a sustainable future.

Inspiration: making vitality visible

Differences between organic and conventional foods have been scientifically proven on various occasions. One process, called sensitive crystallisation, uses copper chloride to create images that reveal the vitality of plants or other substances such as blood. This uncovers clear differences between conventionally, organically and biodynamically produced products. While the structures of conventional crops show tangled, unclear patterns, organically and biodynamically grown plants have clearly recognisable, even beautiful, structures.

Vision Goal 3: Integrative health

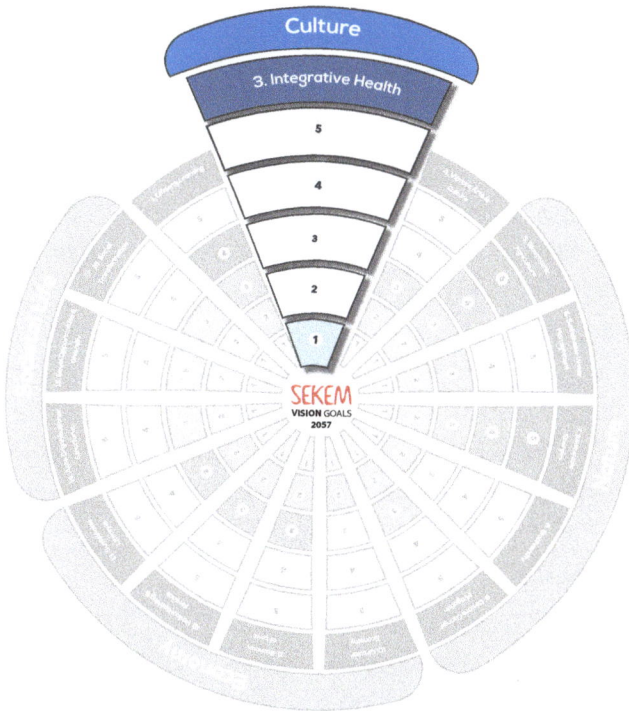

Vision for Egypt 2057
Integrative health and therapy practices are established and widespread throughout the Egyptian healthcare sector.
Progress
The Sekem Medical Centre represents a popular alternative to clinics and medical practices in Egypt. Qualified doctors from different specialisms take their time with their patients and use natural medications and a holistic approach to treatment. The clean, aesthetically pleasing environment contributes to the popularity of the facility, as does its accessibility for people with low financial means. However, the Medical Centre does not yet provide a model for promoting integrative healthcare. To move closer to this goal, a vision group has been set up that has formulated a definition of integrative health. This states that

> having a sense of purpose in life, being surrounded by beauty and feeling connected to a social environment are just as important to an all-round understanding of health and wellbeing as regular exercise, wholesome nutrition and creative self-expression.
> We have 4 ongoing projects and activities, including:
>
> - EcoHealth: holistic health service

We do not want to treat health as a separate, theoretical subject, but as a human good that permeates all spheres of life. We therefore address the various aspects that are necessary for a healthy life and for the prevention of diseases at different levels. In our companies, for example, there is hygiene training, as well as activities to beautify the working environment. Our prevention programmes take into account leisure activities, social factors, nutrition, exercise and stress management. We always try to involve as many stakeholders as possible in a comprehensive way. Our Child Health Project, for example, does not exclusively look at child therapy, but also integrates training for parents and teachers.

The Sekem Medical Centre offers basic medical care for all employees as well as people from the surrounding villages.

At Heliopolis University we have established a faculty for physiotherapy, which we hope will become an exemplary model for holistic, external treatment and disease prevention. For as much as physiotherapy has grown in Egypt in recent years, this branch of education and profession is predominantly chosen for prestige reasons and focuses on treatment with equipment rather than the healing powers that can come from touch. The same is true for pharmacology. Here, too, we want to become an example with our faculty at Heliopolis University and scientifically prove how pharmaceutical work can be done successfully with natural remedies, especially herbal ones.

We have also established a health service that provides health insurance for our co-workers. This programme focuses on preventative measures to maintain health; it also takes into account the self-healing potential of the patient. In this way, we want to reduce the need for conventional pharmaceutical treatments and instead work more preventatively to ensure people's all-around health.

Inspiration: health-promotion programme

Our EcoHealth health service aims to promote holistic health so that fewer illnesses occur in the first place. To this end, all employees pay a 'solidarity contribution' that covers individual counselling and guidance on self-healing practices in addition to supplementary health insurance. The members are accompanied and advised in various areas of life by different experts and specialists, not just doctors and pharmacists but also physiotherapists, nutritionists and psychologists. If their health parameters and general health improve because of the measures, their service contribution decreases. In this way, people are encouraged to take care of their health at an early stage.

Vision Goal 4: Arts and culture

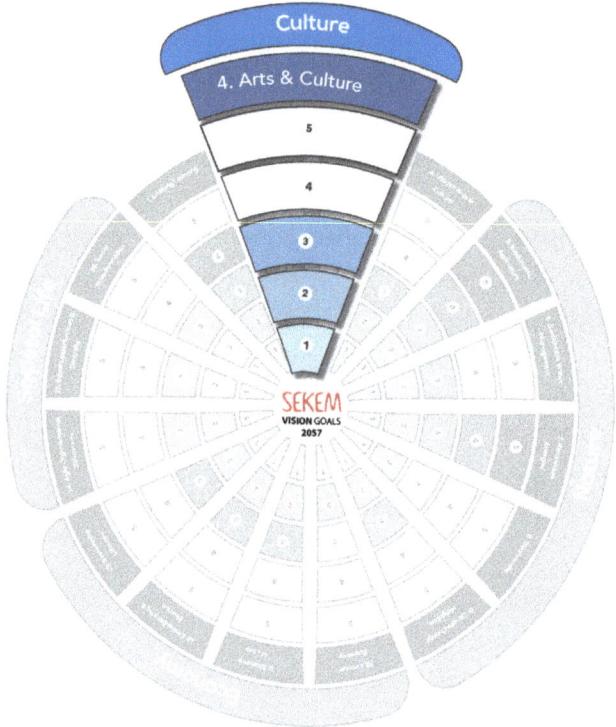

Vision for Egypt 2057
Regional, local, and international arts and cultural activities are valued by and co-created with the Egyptian people.
Progress
We are currently in the model development phase for this goal. We have 3 ongoing projects and activities, including: • art academy at Heliopolis University • three 'Spaces of Culture' in the thirteen Sekem villages

The new amphitheatre on the Wahat farm in the middle of the desert.

We offer many different arts and cultural programmes with the intention of encouraging participation and co-creation. Egypt's existing arts are included, as are contributions from other countries and contexts.

To provide a space for this promotion of culture, we have established three so-called Spaces of Culture:

- a large amphitheatre at the Sekem Mother Farm
- an open-air theatre at Wahat Farm
- the Integral Space of Culture at Heliopolis University

The latter grew out of a vision of Ibrahim Abouleish's called the House of Cultures. This was a cultural hub that gave the various arts a space in which they could be presented and developed. Through the encounter and interplay of diverse cultural expressions a new culture can emerge. Artistic events take place regularly in the Spaces of Culture, which are open not only to members of our community but also to interested people from outside.

At Sekem, employees regularly take part in artistic performances.

Whereas the Space of Culture at Heliopolis University is mainly visited by the urban audience from Cairo, the theatre programme at Wahat Farm is completely different. There, in the desert climate, Bedouins and farmers join the audience and watch as traditional artists present their talents. But here, too, a symbiosis is developing as cultural workers from the city travel to the desert and our students and staff from the Core Program exchange ideas with people from the area.

The concept is now being developed further so that it can enrich many different places throughout the country, from urban Cairo to desert oases and villages along the Nile. We want to test this initially in the thirteen villages around the Sekem Mother Farm. It is important that people have the opportunity to participate in and shape the project so that they feel motivated not only to consume culture, but also to really integrate it into their lives and become creative themselves.

In all of this, we want to promote awareness of the fact that art is not a pastime only for the privileged, rather it unfolds in all aspects of life: in the design of one's home and garden as well as in one's clothing or social behaviour. As the German artist and teacher Joseph Beuys once said: 'Everyone is an artist.'

Inspiration: Space of Culture

The Space of Culture at Heliopolis University includes an amphitheatre, a gallery, a lecture hall and a theatre. These venues regularly host concerts, readings, exhibitions, lectures, theatre performances and similar cultural events. Sometimes these are by Egyptian artists and sometimes by international guests; sometimes they are modern and sometimes traditional. In addition, the staff and students at Heliopolis University get creative by showing and exhibiting their works from the Core Program. There are also regular joint productions by Sekem school pupils with staff and students from Heliopolis University.

6.

Ecology

The deed is the transformation of the earth. The deed on the earth is the attraction for the vision. The vision needs a piece of the earth.

From Ibrahim Abouleish's notebook

Inspirations for an earth fit for our grandchildren

The responsible and sustainable use of nature, and our conscious care and interaction with it, form the basis for all development. Ibrahim Abouleish described this approach in a dream image:

> I carry a vision deep within myself: in the midst of sand and desert I see myself standing at a well drawing water. Carefully I plant trees, herbs and flowers and wet their roots with the precious drops. The cool well water attracts human beings and animals to refresh and quicken themselves. Trees give shade, the land turns green, fragrant flowers bloom, insects, birds and butterflies show their devotion to God, the creator, as if they were citing the first surah of the Quran. The humans, perceiving the hidden praise of God, care for and see all that is created as a reflection of paradise on earth.[1]

In this dream image, inspiration from the Quran is clearly expressed. It is not only about the protection and preservation of nature, but also about its evolution.

In ancient Egypt, care of the land was guided by priests who had a connection to the divine-cosmic order. As a result, people developed a sense of 'right' action on earth. The gods worked through the people, who saw themselves as tools standing between heaven and earth, and they sacrificed part of their harvest to the gods in gratitude. Once, farmers were the caretakers of the land; today they are the producers and beneficiaries. Humanity has little connection to a divine or cosmic sphere. At the same time, religion, the belief in Allah and his holy book the Quran, is of paramount importance in Egyptian culture. The Quran depicts innumerable images of nature and natural processes that often serve as metaphors for divine-spiritual activity, calling for nature to be respected, nurtured and developed. God has granted human beings the ability to view nature in all its phenomena along with the moral responsibility to protect it as a valuable source of life – he offered us this trust (amāna). As God's stewards on earth (ḫalīfa) humans are called to care for the earth responsibly.

These indications from the Quran motivate us to rediscover our lost connection with nature and the cosmos. By observing nature, its richness, diversity and regularities, human beings are directed to divine activity, to the 'one God' in the multiplicity of phenomena, as explained in the Quran:

> In the water that Allah sends down from heaven, and with
> which He revives the earth after its death and lets all kinds
> of creatures spread on it; in the change of the winds and the
> clouds put in service between heaven and earth, in all these
> are signs for people who can know them. (Surah 2:164)

In particular, the balance of diversity in nature, as the ideal of divine creation, is repeatedly described in the Quran:

> And the earth We have spread out. How excellently We have
> levelled it! And of everything We have created a pair, that you
> may consider it. (Surah 51:48–49)

In Surah 50:7–8 the spiritual level is added:

> We have also spread out the earth with mountains fixed upon
> it, and We have caused pairs of every beautiful kind to spring
> forth upon it, for insight and as an indication to everyone
> who devotedly connects himself with the spiritual.

Goethe also aptly gave poetic expression to such an integral view of nature:

> When considering nature's meaning,
> always mind that one is all.
> Nothing outside, nothing inside.
> What inside you see, outside it is.
> So do seize without delay
> what not public mystery must stay!
> Do rejoice in true appearance,

do rejoice in solemn play,
for no living thing is one,
Life is always many.[2]

In Goethe's poetic view we find not only the holistic approach to ecology, which would later be embodied in biodynamic agriculture, but also the cosmic connections in nature to which reference is also made in the Quran. Surah 55:5–9 states:

The sun and the moon run their prescribed course, the stars and the trees incline before the Lord, and the heavens He has lifted up and balanced, you shall not destroy the balance.
Keep the right measure and do not lose it.

By focusing on the interaction of the entire ecosystem, biodynamic agriculture fulfils the human mandate not to unbalance or destroy the created order. Thus, Sekem farmers nurture the vitality of the earth by promoting the balance of minerals and microorganisms, and by working with the different elements of nature. They also take into account the planetary and stellar constellations and thus the interaction between the cosmos and the earth. By incorporating the cosmic and spiritual dimensions into their approach to nature, working with the earth becomes a spiritual act, a 'praise to God'.

As a chemist with a doctorate, Ibrahim Abouleish noted that the symptoms of nutritional deficiency from which people increasingly suffer today can be traced back to the lack of life processes in plants. Although conventional agriculture makes crops grow quickly by using artificial fertilisers, it nevertheless hinders the natural ripening process, causing the plants to lose many of their nutrients and life forces. In the Quran, it is pointed out several times that human beings must consider the ripening of the fruits and not be wasteful:

Eat of their fruits when they are ripe and give a portion of them on the day of harvest, and do not be excessive. He does not love the self-indulgent. (Surah 6:141)

The Quran repeatedly says that we should eat deliberately and only from 'good' sources. Food should be eaten only if it is *ḥalāl* (permissible) and *ṭayyib* (good). In the Quran, these two criteria are usually mentioned together: 'Eat now of that which you have captured, if it is permitted (*ḥalāl*) and pure (*ṭayyib*)' (Surah 8:69).

The question here is why two such similar criteria are necessary. One possible reason could be the physical and moral levels that this refers to. The Arabic word *ṭayyib* can be translated differently depending on the context. As well as meaning 'good' in general, it can also mean 'delicious', 'delectable', 'clean/pure', or even 'ethically correct'. But it is also an intensification of *ḥalāl* (permitted). Nowadays, *ḥalāl* usually refers exclusively to the method of slaughter. In the Quran, however, its original meaning goes far beyond this. When it is said, 'Eat of the good (*ṭayyib*) that we have bestowed upon you, but not without measure' (Surah 20:81), it becomes clear that when consuming food, attention should always be paid to good quality, that is, its provenance. Moreover, food should not be wasted, and the natural balance and harmony of nature should not be disturbed. In today's context, we understand this as a mandate for sustainable agriculture that does not use pesticides, growth accelerators or genetic engineering, all of which interferes with the natural balance.

At Sekem, we recently weighed all the leftover food that was returned to the kitchen after the communal lunch. When we told the community how many more people could have been fed from these leftovers, we were ashamed. But it helped make this problem easier to grasp so that a better way of dealing with it can be found.

The Quran also provides information about animal husbandry. Animals should be included in the natural cycle of agricultural activity:

> And He created the animals for you. They provide you with warmth and other benefits, and you eat from them. And you delight in their beauty when you drive them in at night and drive them out in the morning. (Surah 16:5)

Like human beings, they are God's creatures and should be treated with appropriate respect:

And there is no animal that walks on the earth, nor any bird
that flies with its wings, but are communities (*ummah*) like
you. (Surah 6:38)

Biodynamic agriculture offers a species-appropriate approach to
animal husbandry as well. For example, emphasis is placed on not
dehorning the cows as this causes a strong change in the nature of the
animal and affects the quality of the milk. The danger of such changes
is addressed in the Quran when Satan says:

And I will mislead them, and will lead them astray to desire
and delusion, and I will command them that they clip the
ears of the herd animals, and I will command them that they
change the creation of Allah. (Surah 4:119)

Such indications inevitably make us think of genetic engineering.

A closer look at Quranic statements concerning the quality of
food makes it clear that the consumption of products that have been
alienated from the natural and divine context is in principle *ḥarām*,
meaning forbidden. The Quran acknowledges the divine context
by giving thanks to God. If the production of food has been done
ethically, then the so-called *basmala* – the invocation formula, 'In the
name of the merciful and gracious God' – is still spoken to give thanks
to God. Thus, in accordance with Surah 5:4, it is declared that the food
is produced in harmonious balance with nature, which is connected
to cosmic and spiritual spheres: 'eat of what they catch for you and
mention the name of Allah over it.'

Surah 2:172 states:

O you who believe, eat of the best things we have provided
for you and give thanks to Allah, so serve Him.

Giving thanks as an acknowledgement of God and an all-embracing
world order, which understands nature in its wholeness and in its
relationship to the divine, is similarly described in a verse by Rudolf
Steiner:

The light of the sun is flooding
The realms of space;
The song of birds resounds
Through fields of air;
The tender plants spring forth
From Mother Earth
And human souls rise up
With grateful hearts
To all the spirits of the world.[3]

Seeing more than pure ecology in nature and recognising in agriculture a cultural work guided by human insight are foundational to Islam. They have found their practical implementation at Sekem through biodynamic agriculture. We have made it our task to bring the connection between nature and its Creator more strongly into people's awareness so that they realise their role and the responsibility connected with it. Goethe commented on this when he said:

> Nature understands no jesting; she is always true, always serious, always severe; she is always right, and the errors and faults are always those of humanity.[4]

For Ibrahim Abouleish it all began with this nature-culture work, and it still forms the basis for the Sekem vision, because all other dimensions are rooted in this spiritual and holistic view of nature.

Forty years of cultivating the desert

Today, avenues of Casuarina trees, palm groves and flowering fields of chamomile, calendula, peppermint and other medicinal plants stretch across what was once desert ground. Birds of various species chirp, sheep graze and cows enjoy fresh clover in the shade of their stalls. In the early morning hours, fog still lies over the Sekem Mother Farm, while the first farmers begin to irrigate and cultivate the fields.

Sekem works with several hundred organic farmers throughout Egypt.

Within forty years, more than 2,000 hectares (4,950 acres) of land have been sustainably reclaimed and cultivated. In the process, hundreds of thousands of trees have been planted, which, together with the now fertile soil, have sequestered many thousands of tonnes of CO_2.[5] Biodynamic agriculture has also brought other benefits for the local ecology and for people in Egypt. Sekem produces compost through a completely sustainable process that contributes to the rich fertility of the soil. In combination with the biodynamic preparations, the soil of the formerly barren desert floor stores 20–40% more water than conventionally cultivated land. We call compost the farmer's 'black gold' because it is used not only to reclaim desert land but also as a fertiliser. It is ideal for dealing with the country's hot, dry climate. Even conventional farmers across Egypt have been persuaded by its many benefits; they have begun to produce and sell biodynamic compost themselves. Meanwhile, the word has been adopted into Arabic, '*kumbust*', and is used in rural areas to advertise this natural fertiliser. It has become a substitute for the Nile mud that used to make the land fertile and has been absent since the construction of the Aswan Dam in the 1960s.

The use of biodynamic-produced compost to reclaim and fertilise the soil shines forth as the most successful initiative Sekem has brought to Egypt. It has become well known throughout the country. The compost's effectiveness is evident without needing to share visions and long explanations; its production and use are easy and inexpensive for anyone to implement. We aim to achieve the same success with our other vision goals.

Today, around five hundred Egyptian farmers who practise biodynamic agriculture have long-term contracts to grow produce for Sekem. This guarantees them stable prices and they receive regular training and education. In addition to technical knowledge, we also offer the farmers and their families the opportunity to take part in educational programmes, literacy courses or training on health topics among other things.

The Sekem Mother Farm, which covers around 100 hectares (250 acres) of land, operates its own wastewater-management system. All wastewater is recycled and used to irrigate the trees. There are several solar-powered systems that provide renewable electricity to the buildings. Due to reduced CO_2 emissions and sustainable agriculture, Sekem has not only been carbon neutral but climate positive for many years.

After demand for our products, both in Egypt and internationally, grew in the early 2000s, and urban pollution increased dramatically, we decided to buy three more land parcels: in Egypt's western desert, in the government district of Al-Minya and in the Sinai Peninsula. With a total of around 2,000 hectares (4,950 acres) of desert, we intend to support not only agricultural activities but also the establishment of rural communities (see Vision Goal 16: Social transformation). However, during the time of our planned expansion the revolution in Egypt severely limited its implementation. Although expansion has progressed slowly since then, olives now grow in Sinai and herbs sprout in Minya. On the farm in the Wahat El-Bahareyya oasis, jojoba, date palms and other trees comprise the main crops. Ecological agricultural research is also being carried out there.

The greening of the desert helps to create a fertile foundation for human development.

Challenges and opportunities

Egypt is known worldwide as one of the oldest civilisations to have practised agriculture. Thousands of years ago, annual Nile floods left fertile soil for people to grow their food. Since it rarely rains in Egypt, the Nile was like a well in the desert that fed the population. But times have changed. The construction of the Aswan Dam in the 1960s meant that the annual floods that brought fertile mud ceased. As a result, more and more chemical fertilisers and pesticides have been used to increase crop yields. Much of the energy generated by the Aswan Dam has been used to produce artificial fertilisers. In the 1990s, Egypt was one of the world's largest consumers of chemical pesticides. With this development, the demand for water also increased, and this in a country where agriculture accounts for more than 84% of all water consumption. According to the United Nations, Egypt has been living in water scarcity for some time now and is well below the water-poverty line.[6] As a result, the country has become heavily dependent on food imports.

To make matters worse, more and more water is being diverted from the Nile, which leads to greater intrusion of seawater and the salinisation of the Nile Delta. This, along with sea-level rise, contributes to making Egypt one of the countries most affected by climate change in terms of rising sea levels.

Egypt also suffers from the loss of fertile farmland as enormous population growth leads to increasing construction on arable land. Other major challenges include environmental pollution from pesticides and artificial fertilisers, inadequate waste disposal, and CO_2 emissions. Only 60% of waste is collected in Egypt and only 20% of that is disposed of properly.[7] Only larger cities have landfills; most of the waste, especially in rural areas, gets burned illegally or even dumped in the desert. Cairo is one of the most polluted metropolitan regions around the globe, with one of the highest levels of particulate matter. More than 90% of the electricity, which was heavily subsidised in the past, comes from fossil fuels. As the population grows, the price of fossil fuels increases and resources become increasingly scarce, not to mention the dramatic corresponding loss of biodiversity.

Numerous studies predict that all the problems that Egypt faces today will increase in the coming decades; furthermore, they will become challenges facing the rest of the world, too. Insufficient water, lack of fertile land and biodiversity loss usually emerge as the most serious challenges in future forecasts. Whereas in the middle of the last century there were just under 0.5 hectares (1.25 acres) of fertile land for every person worldwide, today there are only around 0.19 hectares (0.5 acres), and by 2050, with still burgeoning populations, this figure could fall to around 0.16 hectares (0.35 acres) per capita.[8] In Egypt today, only 400 square meters are available for each inhabitant. The increased demand for food (by around 60%, according to FAO estimates) will result in an increased demand for water. Worldwide, agriculture is the largest consumer of water.

Sekem wants to counter these negative forecasts with what we have achieved since 1977 through a holistic and sustainable approach to nature and agriculture. The balance of Egypt's ecosystem must be restored as soon as possible if it is to be viable.

Vision Goal 5: Agriculture

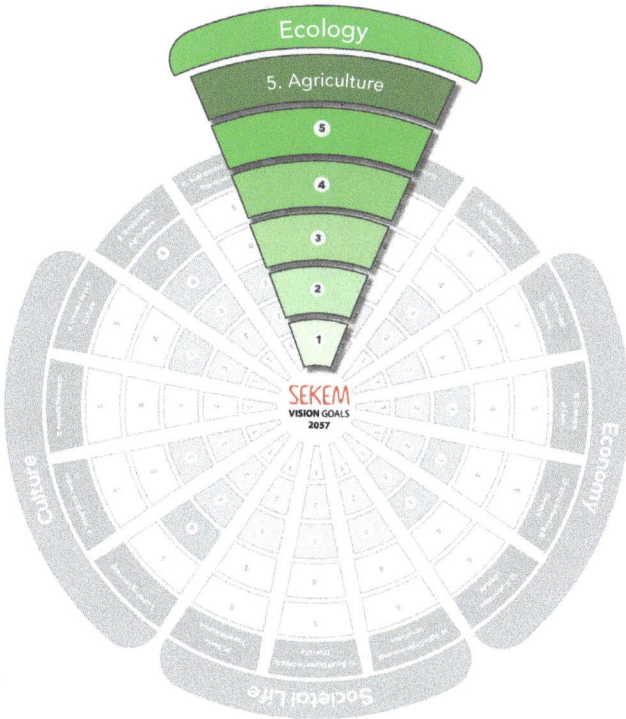

Vision for Egypt 2057
Egyptian farmers are implementing sustainable agriculture using biodynamic or organic methods.
Progress
We are already at the stage with some projects where we can achieve nationwide relevance and real change can occur. We have 21 ongoing projects and activities, including: • technical support for biodynamic and organic farmers • adapting crops to climatic conditions • 'Greening the Desert' project at Wahat Farm

Sekem wants to support the seven million farmers in Egypt in switching to sustainable agriculture.

Egypt's first Faculty of Organic Agriculture opened at Heliopolis University in 2018. Students receive both scientific theory in the classroom and extensive hands-on experience on the Sekem farms. Several courses are conducted in cooperation with partners, for example, with the Agricultural Section of the Goetheanum in Dornach, Switzerland. In this way, we hope that our faculty's teaching will become a model that can be emulated across Egypt and beyond.

We also conduct research using various methods to visualise the quality and vitality of organic and biodynamic foods (see page 66 Inspiration: making vitality visible). We are establishing a dedicated laboratory at Heliopolis University that will allow students and scientific researchers to prove the benefits of organic food. This will hopefully expand the market for organic food and at the same time advance our vision for the production of more healthy food in Egypt.

We can prove that organic agriculture, and especially biodynamic agriculture, offers solutions to the many major environmental challenges facing us today. For example, the ongoing study 'The Future of Agriculture in Egypt' conducted by Sekem and Heliopolis University shows that for the five most important Egyptian crops – cotton, corn, potatoes, rice and wheat – organic farming is already more cost-effective than conventional farming. The study employs what is called true cost accounting, a holistic approach that considers factors such as soil, air and water pollution, which, sooner or later, lead to higher costs in conventional agriculture. With research applications such as these, we work to create awareness of the benefits of sustainable agriculture at various levels. This in turn helps policymakers and civilians make better decisions regarding food production and consumption. Because this kind of research more accurately reflects the reality of the future, change will occur sooner as a result.

Sekem, Heliopolis University and our international partners collaborate to secure funding for projects in the organic sector in Egypt and other African countries. One example is the 'Knowledge Hub for Organic Agriculture in North Africa'.[9] Here, in cooperation with partners in Tunisia, Morocco and other African countries, knowledge on organic agriculture is being collected, processed and made widely available.

An important topic regarding organic agriculture in Egypt is that of seeds. We are increasingly breeding our own biodynamic seeds and making them available on the Egyptian market. We research varieties suitable for North Africa and the Mediterranean region, and Sekem farmers test these before they are put on the market. Sekem seeds are also available to amateur gardeners who implement urban gardening projects on Cairo's rooftops. The aim is to offer a diversity of graded seeds to build alternatives to the multinational corporations that increasingly monopolise the seed market, and thus global food production, with patented hybrid seeds.

To support farmers who want to convert their farms, Sekem has founded the Egyptian Biodynamic Association (EBDA). In close connection with the Biodynamic Federation Demeter International (BFDI; formerly known as Demeter International), EBDA's capacities are being expanded to serve the growing number of biodynamic farmers in Egypt, who are among the most important multipliers for change. We have already reached several thousand farmers and are confident that by 2027 we will have reached a critical mass of Egypt's seven million farmers.

A visible change will then occur in agriculture – from the niche to the mainstream.

In our 'Greening the Desert' project on Wahat Farm (see Vision Goal 16: Social transformation), we are testing how large-scale desert greening using organic farming techniques on arid, desert-like soils are more economical and productive than those of our conventional neighbours. One factor here is the revenue generated by CO_2 sequestration in soil and trees (see page 103, Inspiration: a positive carbon footprint in the desert).

Together with the BFDI, we are developing a strategy that will soon allow all Demeter farmers to determine their CO_2 sequestration and receive financial compensation for it. This will help make the conversion to biodynamic agriculture more attractive and, accordingly, more CO_2 will be sequestered in agriculture not only in Egypt but also worldwide (see also Vision Goal 9: Climate neutrality).

Farmers making the biodynamic preparations that play an important part in the sustainable reclamation of the desert.

However, our vision does not only point to organic or biodynamic agriculture as a solution. We believe that organic criteria alone will not suffice for a truly sustainable approach to nature. Pesticide limits, the renunciation of chemical fertilisation and the prohibition of genetically modified seeds are important basics, but sustainable agriculture involves much more than that. For example, the humus content in the soil, the use of the right seeds, CO_2 sequestration, but also social standards or development opportunities for people on the land are important in order to act in a truly sustainable way (see also Vision Goal 11: The economy of love).

In addition to research and practical projects, Sekem and Heliopolis University want to promote knowledge of sustainable development and chemical-free farming. For some time Sekem schools have taught these topics as part of their natural resources' curriculum. Now they seek to spread this understanding to other schools in Egypt. Information sharing also occurs through Sekem's organic company, Isis Organic, as well as the Sekem brand itself. At the same time, Sekem farmers are trained to share their experiences with farmers who use conventional methods.

Inspiration: biodynamic agriculture for 17,000 smallholders

By the end of 2024 we had over 17,000 farmers in Egypt trained in biodynamic farming methods and converting from conventional agriculture.[10] Since not all of them can be contracted directly by Sekem, we provide support in the purchase and marketing of their products, especially during the three-year conversion period, when these products cannot yet be certified. We also offer to certify their CO_2 sequestration. The revenue they receive from us adds to their income.

The current number of biodynamic farmers is only a small percentage of the country's seven million farmers, but we are confident that we will reach 40,000 farmers by the end of 2025 and soon achieve the critical mass necessary to allow the expansion to proceed on a large scale in the coming years.

Vision Goal 6: Sustainable water management

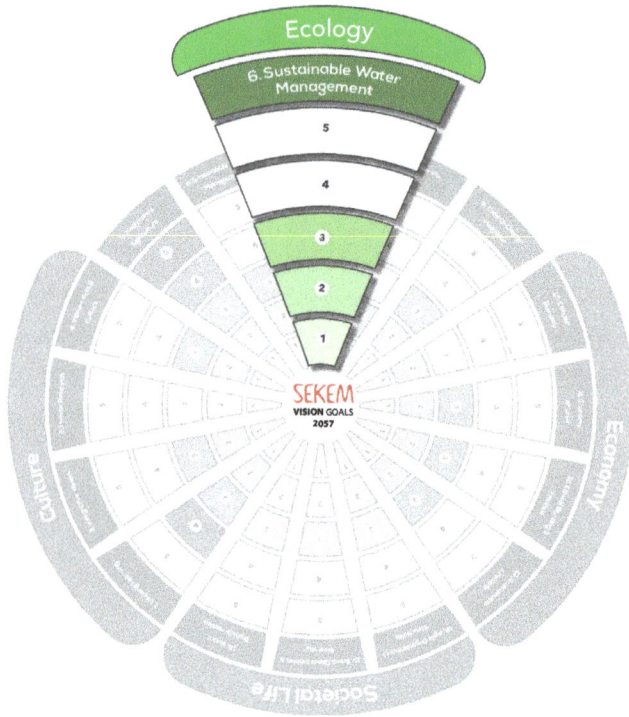

Vision for Egypt 2057
Egypt operates a sustainable water-management system that optimises water consumption, reuses wastewater, and uses innovative water-harvesting and irrigation techniques.
Progress
Given the major problem of water poverty in Egypt, this vision goal is particularly close to our hearts, even if it does sound a little utopian. Our organically managed soils allow us to use 20–40% less water than conventional farmers, but Sekem still needs to optimise its own water use, especially in agriculture, which consumes most of the water in Egypt. To do this, we must further develop existing irrigation methods and research alternative systems.

We will also operate all irrigation systems with renewable energy.
We have 5 ongoing projects and activities, including:
- seawater desalination
- decentralised wastewater treatment

We have been implementing the purification and 100% reuse of wastewater for a long time. But this is only sustainable if the wastewater is not already contaminated with harmful chemicals. After all, agricultural water in Egypt is used up to three times and additional harmful substances are released into nature through the residues from conventional agriculture. Organic agriculture prevents this.

To meet the increasing demand for water in Egypt, alternative irrigation techniques need to be researched and methods that so far have only been worked out in theory need to be made practicable. One such model is seawater desalination. We want to take a closer look at this and see if it can work without the high use of fossil energy. We also need to find out how the separated salt can be used so that the

technique can help develop a sustainable and inexpensive source for Egypt's water management.

There are trees and plants that require almost no irrigation and get their liquid mainly from the moisture in the air. Some of these trees and plants are already being tested in the 'Greening the Desert' project (for example, moringa or acacia). We are seeking and researching other innovative technologies and solutions.

Inspiration: decentralised wastewater treatment

Currently, only about 15% of wastewater in Egypt is recycled. In the thirteen villages around Sekem Farm, we are testing a so-called decentralised wastewater treatment tower. The system purifies wastewater so that it can then be reused for agricultural irrigation without harming the environment. The module is built against the walls of houses so that it saves as much space as possible. It holds great potential for smallholder farmers to greatly reduce agricultural water use.

Vision Goal 7: Renewable energy

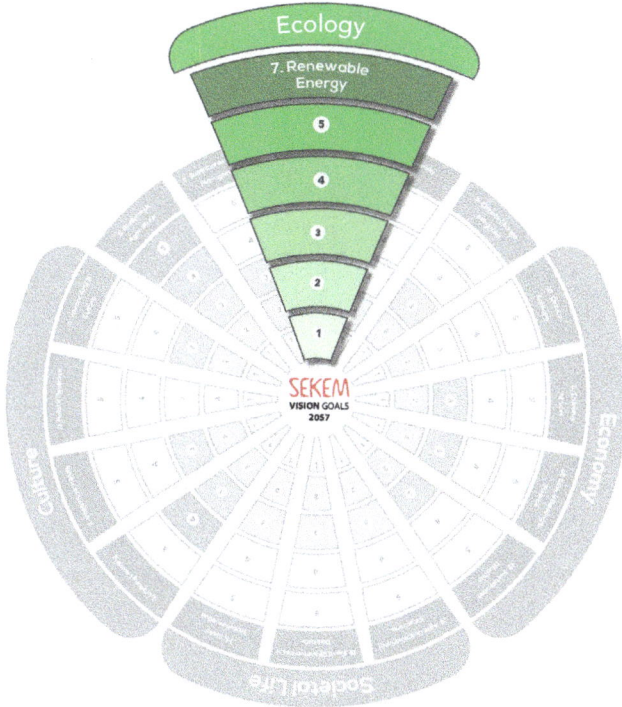

Vision for Egypt 2057
Sustainable energy management based on renewable energy and optimised consumption is being implemented in Egypt.
Progress
The goal of making renewable energy production a mainstream method in Egypt has already been advanced to the point where we are in the phase of the widespread rollout of tried and tested models. We have 12 ongoing projects and activities, including: • solar-powered water-pump system   • energy concepts for the national power grid

Trainees from the Sekem Vocational Training Centre install and maintain the solar equipment, and students from Heliopolis University participate in the research.

The realisation of this vision might not seem difficult when you consider the great potential Egypt has for developing renewable energy: the sun shines almost all year round and there are extensive coasts with plenty of wind. For a long time, however, the development of these alternative energy sources did not progress very far, due largely to the cost of installation, the difficulty of maintaining equipment, and the large subsidies received by energy companies that made fossil fuels extremely cheap. Another obstacle is the lack of expertise in both the installation and subsequent maintenance of solar and wind systems.

We started our first experiments in alternative energy many years ago with photovoltaic systems, wind turbines and hybrid solar dryers for herbs. There are now several large photovoltaic systems on the Wahat farm in Egypt's western desert. These are mainly used to operate the irrigation systems, but also supply energy to residential buildings.

In the desert, wind is a good source of renewable energy in addition to the sun. We are conducting research into this with a wind turbine.

Efforts are also constantly being made at the Sekem Mother Farm and Heliopolis University to promote the use of solar energy, especially through research and professional training. At Sekem's Vocational Training Centre, apprentices can specialise in solar-thermal and photovoltaic technologies during their training, and at Heliopolis University, engineering students and lecturers, some from other Egyptian universities, take part in specific advanced training programmes.

In this way, we have optimised our goal in this area and aim to eventually run all of Sekem's institutions on 100% renewable energy. We managed to achieve this for Heliopolis University by the end of 2021.

Our energy company EcoEnergy is licensed to design and implement renewable energy ideas and connect them to the national power grid. The company then sells the green electricity to Sekem facilities. In this way, EcoEnergy could also supply the surrounding thirteen villages with renewable energy in the longer term. The goal is for this electricity to be priced the same as conventionally generated electricity, ideally even cheaper.

Another step is e-mobility. With our sustainably generated electricity, we can already power some of the electric vehicles at Sekem, and by 2027 we aim to power all our vehicles by renewable energy.

In recent years, a new awareness seems to have emerged regarding the potential of renewable energy in Egypt. More and more plants are being set up that generate both solar and wind energy. This is because, in Egypt, solar energy is cheaper than energy produced by diesel generators when there is no access to the national grid. Alongside our ongoing research, we see it as our task to impart the necessary skills and training to our co-workers so that they can reliably install and maintain the equipment.

Inspiration: regenerative energy generation

On Wahat Farm we use a pivot-irrigation system that uses sprinklers to water crops in a circular pattern around a central pivot point. This is now powered entirely by renewable energy. We carried out a lot of research not only to convert the water pumps from diesel generators to renewable energy, but also to power the mechanised pivot system itself via solar energy. These systems are designed by the lecturers and students of the Faculty of Engineering at Heliopolis University, and trainees from the Sekem Vocational Training Centre take care of installation, commissioning and maintenance.

We have also developed small solar-powered systems for our farmers. With 5-kW water pumps they can replace their diesel generators for irrigation, and with a 2-kW model they can supply their house with energy. Sustainable energy generation does not cost them more than the conventional alternatives, and our microcredit programme provides the funding.

Solar panel array at Wahat Farm.

Vision Goal 8: Biodiversity

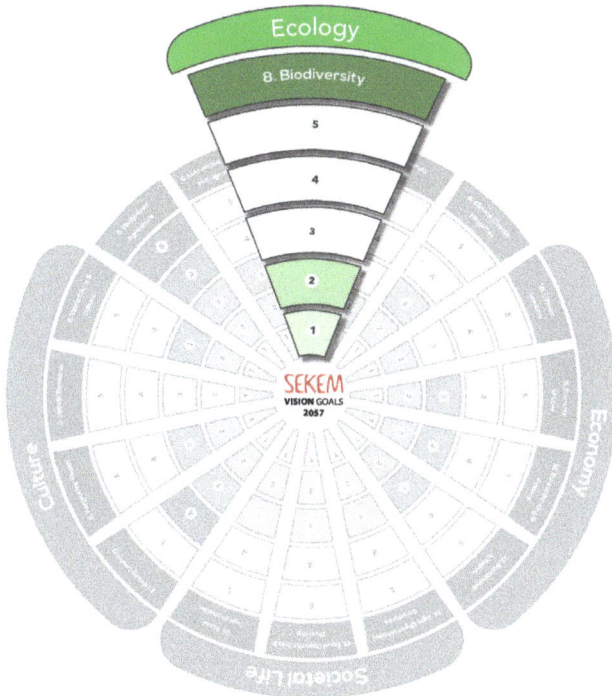

Vision for Egypt 2057

Egypt's biodiversity is sustainable, growing and flourishing.

Progress

Protecting natural ecosystems is important for maintaining biodiversity, but that alone is not enough. Half of the habitable land on earth consists of agricultural land cultivated by humans. As a result, biodiversity tends to decline. So, it is extremely important to redesign these agricultural systems to promote diversity. We have 8 ongoing projects and activities, including:

- using biodiversity levels as a measure of the economy
- observing and compiling an inventory of existing species at Wahat Farm solar-powered water-pump system

*The Egyptian bee (*Apis mellifera lamarckii*) is just one example of Egypt's endangered species.*

On average, organically farmed fields have one-third more diverse species and twice as many organisms as an area of the same size that is conventionally farmed. The fertile soil structure and diverse crop rotations in organic farming contribute to this greater biodiversity. The proportion of hedgerows or pristine meadows, which are mandatory in biodynamic agriculture, also create habitats for greater biodiversity.

At Sekem we practise agroforestry, which is the cultivation of trees and shrubs as part of our farming system. Planting trees is therefore an important part of our understanding of sustainability, and by 2027 we want to plant one million more trees, which will benefit biodiversity. We also support our farmers in planting trees by integrating them into our carbon offset programme. In this way, every tree is financed, and, as a result, biodiversity increases.

Besides these contributions, we want to work specifically for the conservation of certain species. One important project in this context is the protection of the Egyptian bee (*Apis mellifera lamarckii*) which is almost extinct. This bee, probably one of the oldest species of bee, has been displaced by highly bred European bees, and its extinction would have drastic effects on the Egypt's entire ecosystem. Sekem is now home to one hundred hives for Egyptian bees and promotes their conservation in many ways, for example, by training beekeepers.

We are also incorporating the effects of biodiversity in our financial statements and research. This highlights the benefits of biodiversity and the costs of its loss in the economic figures.

Inspiration: monitoring biodiversity in the desert

In order to conserve biodiversity it is necessary to monitor and compile an inventory of the species currently present. One of the assumptions behind the transformation of the desert ecosystem into an agroecosystem is that it will have a positive impact on the species living in that area. We therefore study the biodiversity of Wahat Farm by first determining the population of a species and then monitoring it. Such research will become an integral part of the agriculture curriculum at Heliopolis University so that we can ensure that our agriculture promotes rather than harms biodiversity.

Vision Goal 9: Climate neutrality

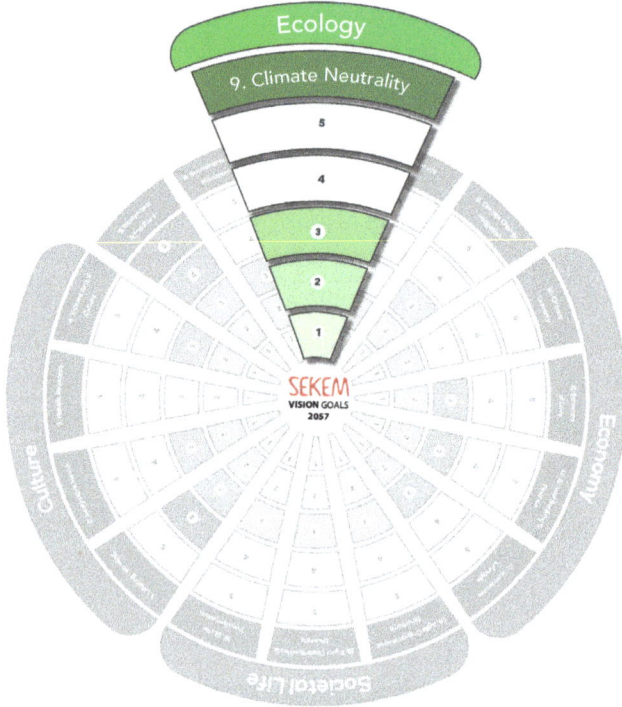

Vision for Egypt 2057

Egypt is climate neutral because no CO_2 is emitted that nature cannot reabsorb.

Progress

We want to make a significant contribution to Egypt achieving climate neutrality. By 2027, we want to plant one million trees, which will bind 33,000 tonnes of CO_2 annually. But the trees only make a relatively small contribution. Several other of our projects and activities mentioned earlier, especially sustainable agriculture, offer solutions for combating climate change. We have 9 ongoing projects and activities, including:

- the creation of high-quality CO_2 certificates and the establishment of emissions trading

Biodynamic agriculture makes an important contribution to combating global warming.

Industrial agriculture and forestry have been the largest contributors to climate change in the last 200 years. According to the Intergovernmental Panel on Climate Change's 2014 *Climate Change Report*, they are responsible for almost 24% of global greenhouse gas emissions.[11] Conversely, this means that sustainable agriculture offers one of the best solutions for mitigating climate change.

Let's take Egypt as an example. In 2020, carbon emissions were around 300 million tonnes. Research on our own farms has shown that organic soils sequester the equivalent of more than 3 tonnes of CO_2 per hectare per year. So if the entire agricultural area of the country (around 3.8 million hectares or 9.4 million acres) were farmed organically, that would save more than 11 million tonnes, compared to the 32 million tonnes produced by conventional agriculture that year.

Globally, it's a similar picture. According to the IPCC, the land used for agriculture worldwide (1.6 billion hectares or 3.9 billion acres), would, if it were managed sustainably, have the potential to sequester

8.6 gigatonnes (8.6 billion tonnes) of CO_2 a year, which amounts to 5 tonnes of CO_2 per hectare per year.[12] If we assume only the minimum value we calculated of 3 tonnes of CO_2 per hectare per year, then the 1.6 billion hectares would save just under 5 gigatonnes of CO_2 through organic management.

Planting trees offers further potential. For a healthy agricultural cycle when growing in the desert, it makes sense to plant about 220 trees per hectare (2.5 acres), which together sequester an average of 5.5 tonnes of CO_2 per year. If biodynamic agriculture was adopted more widely, this would amount to an additional 8.8 gigatonnes of CO_2 savings. And these figures do not include the potential of natural forests; only trees cultivated on agricultural land were taken into account.

In addition, the elimination of artificial fertilisers and pesticides in organic and biodynamic agriculture helps mitigate climate change. The production of nitrogen fertiliser alone consumes an extremely large amount of energy. This causes up to 0.6 gigatonnes of CO_2 per year worldwide, not to mention the high nitrous oxide emissions (up to 300 times more potent and more damaging to the climate than CO_2) that result from nitrogen fertilisation and cause severe damage to groundwater and the air.

These savings from a shift to biodynamic agriculture – 5 gigatonnes from soil sequestration, 8.8 gigatonnes from trees, and 0.6 gigatonnes from the elimination of pesticides – would prevent about 14.5 gigatonnes of CO_2 of global greenhouse gas emissions – more than one-third of global emissions. By comparison, in 2019, global CO_2 emissions were 36.4 gigatonnes.

Two other aspects that are important to us in combating climate change are research and communication, or rather our work in education. Heliopolis University and the affiliated Carbon Footprint Centre identify the sources of greenhouse gas emissions and research innovative solutions. Awareness is raised, education is provided, and know-how is disseminated through teaching students, as well as through the training of teaching staff from all over Egypt. The Carbon Footprint Centre manages carbon offsets for Sekem as well as other companies in the country and internationally.

Recently, we and all our co-workers calculated our individual footprints and estimated how many earths we would need to maintain our standard of living. The result was shocking. Although none of us live in luxury, an average of 2.5 earths was the sobering result. We are working to improve the underlying causes and will repeat this survey next year.

Inspiration: a positive carbon footprint in the desert

Sekem has been climate positive for many years. On our desert farm in Wahat, we can now show how this is possible through sustainable agriculture and how this can neutralise the CO_2 footprint of many people at the same time. In 2021, we were able to sequester almost 14,000 tonnes of CO_2 (equivalent to the carbon emissions of 5,600 Egyptians) by binding CO_2 in the trees and soils and using compost and renewable energies. With the expansion of agricultural activities on the farm, this will bring us to over 69,000 tonnes in 2027. According to current figures, this would correspond to the carbon emissions of around 27,000 Egyptians. By way of comparison, around 40,000 people live in the Wahat El-Bahareyya oasis, where our farm is located.

7.

The Economy

When numbers and figures no longer
Are the keys to all creatures,
When those that sing or kiss,
Know more than the learned
...
And in fairy tales and poems
One sees the true histories of the world,
Then the whole inverted being
Flies away before one secret word.

'When Numbers and Figures'
Novalis

Inspirations for an economy of love

The economy is as significant a part of the Sekem vision as the culture, environment and social life. Sekem's economic activity enhances and supports the activity of the other spheres. Conversely, economic success depends on good raw materials, legal structures as well as human potential and intellectual property – the economy needs the impulses from all spheres and at the same time makes them possible in the first place. Several of today's problems seem to result from the fact that in wider society the economic sphere dominates all others. To address this, Sekem prioritises personal growth and development to ensure a successful economy, shaping the economy in such a way that it serves the development of people rather than the other way around. The economy should benefit the people who produce products and services, such as by paying them a fair wage for their work that enables them to lead meaningful lives, and the products and services should, in turn, have a positive impact on consumers. This most original sense of economic activity seems to have been completely eclipsed in today's world. Anonymity, shareholder value, consumption, monopolisation, the interests of individuals and ever-increasing inequalities define today's economic life.

The situation in which we find ourselves at the moment can be better understood if we consider humanity's current state of consciousness (see Chapter 9: Our Vision for Egypt 2057 in the Context of Current Events). In the prevailing consciousness, broadly speaking, people's interests usually start with themselves. Wealth and status assume great importance and blinds us to what else might lie beyond their scope. This is not ill will, but a lack of awareness of the bigger picture.

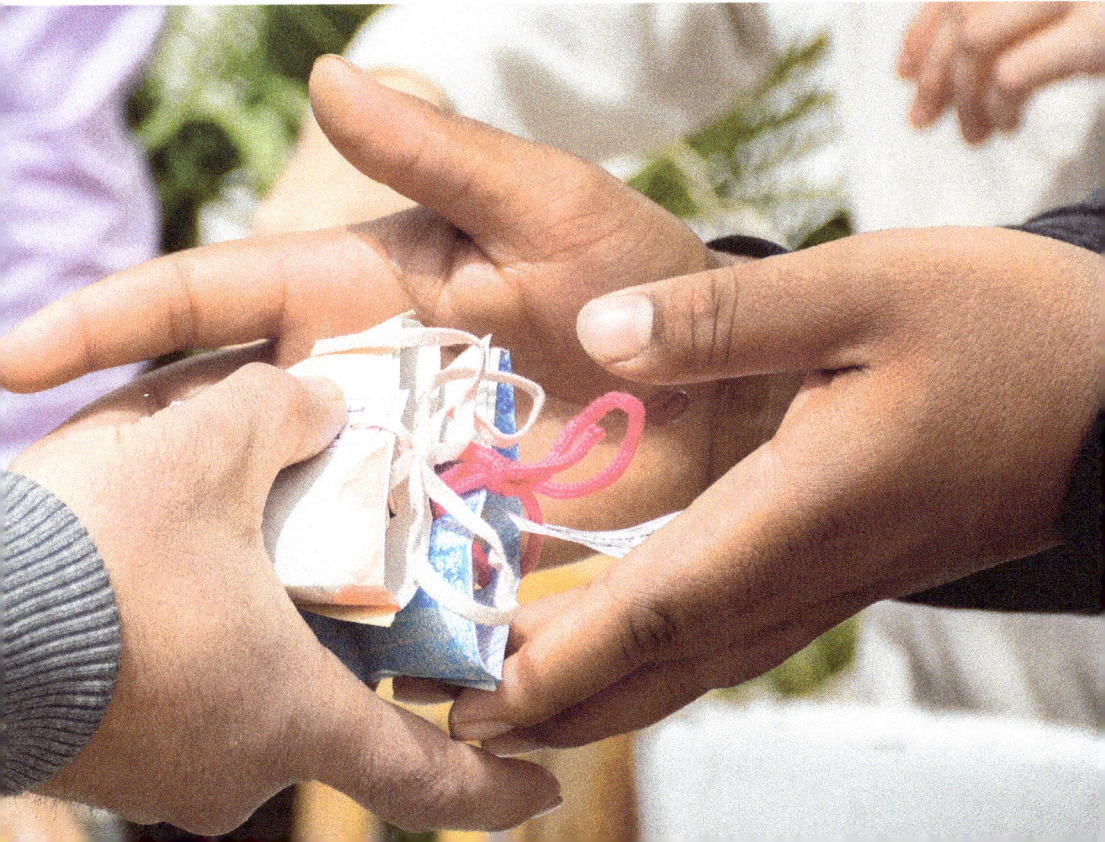

To give the economy its proper place in society so that it does not dominate the other spheres of life is part of the Sekem vision, one that is shared in their own way with many scholars and researchers from the East as well as the West.

Ibrahim Abouleish called the Sekem economic model an 'economy of love', and he took the central message of love from Islam and Christianity in order to demonstrate the importance of acts of compassion in the economy. Again and again in Islam, as in other world religions, love and the deeds arising from it are presented as the highest ideal that no worldly wealth can come close to. The Quran states:

> Wealth and children are the ornaments of this worldly life. But what remains, the good works – they bring with your Lord a better reward and establish a better hope. (Surah 18:46)

And even more clearly:

> You will not obtain love until you give of that which you love. (Surah 3:92)

In addition to this central message, which is the basis for implementing an economy of love in Sekem, other essential values for an economic model based on solidarity are also rooted in Islam, such as gratitude, humility, fraternity and social responsibility. Surah 16:90, for example, describes the importance of justice:

> Allah commands to do justice, to do good and to give gifts to the neighbour. He forbids the shameful, the reprehensible and the violent.

If we take these values seriously, our economy-of-love concept is a logical and consistent consequence.

In today's increasingly complex and abstract globalised economy, any direct contact with how something is produced – and therefore with anything ethical – has been lost. Ibrahim Abouleish found this reference and the basic idea of an economy of love especially in Rudolf Steiner's concept of associative economics:

> When travelling through Italy visiting biodynamic farms
> ... before coming to Egypt, I soon noticed that one of the
> most important prerequisites for successful economics was
> completely lacking: the awareness for associations.[1]

Associative management involves a union of consumers, traders and producers all working together to produce meaningful products and providing participants with a fair share of the proceeds. Associations are governing bodies in economic life that, starting with the need of the consumer, determine a fair share for all parties involved in production. And all parties involved in value creation and consumption are part of such an association. As Ibrahim Abouleish wrote in his autobiography:

> An association is founded on an agreement that gives security
> to all involved. The basis of an association is thus mutual
> trust or, in other words: economy based on fraternity.[2]

An Islamic Hadith states in this regard: 'None of you is a believer unless he wishes for his brother what he wishes for himself.'[3]

An initial example of such an associative, sisterly approach is the network of farmers, businesses and Sekem trading partnerships. Our partners in Europe, for example, are part of associations that promote similar values, a form of an economy of love.

If we look further into the Quran, there are also references to the individual criteria for an economy of love, which we have either taken up directly or adapted in Sekem. We also promote them with our vision for Egypt 2057. One very important idea is that money should never be accumulated but should always be allowed to circulate so that it can have a meaningful effect on people. Thus, the Quran states:

> O you who believe, verily, many of the scribes and monks
> consume people's wealth unjustly and turn away from Allah's
> way. And to those who hoard gold and silver and do not use it

for the way of Allah – they are promised painful punishment.
(Surah 9:34)

It is also reported in a Hadith how the Prophet Muhammad said:

Do not close your money bag, otherwise Allah will also close
His blessings from you. Spend in Allah's way as much as you
can.[4]

Such statements not only underpin the concept of *zakāt*, the
obligatory religious tax, but also run parallel to thoughts developed by
Rudolf Steiner that are important for a sustainable economic model:

Capital must be used up, until what remains we may conceive
of as a kind of seed to kindle the economic process anew –
once more from the starting point of nature.[5]

The economy of love does not in any way prohibit making a profit,
rather it is a matter of ensuring that the money does not stand idle but
is invested in the public sphere: in cultural life, for example, or in the
protection of nature. We find this thought also in the Quran:

And who, when they donate, are neither wasteful, nor
restrained, but keep the middle in between. (Surah 25:67)

In addition, at Sekem, we strive to ensure that the companies that
make a profit pay a fair price to the agricultural sector. After all, in
today's global economic system with distorted market prices, the
agricultural sector is no longer able to support itself.

We do not view success or economic competitiveness as negative, but
rather as necessary and valuable contributions to the application and
promotion of intellectual achievement. What is always of importance,
however, is how to deal with the added value generated.

Again and again in the Quran there are statements about the
sharing of wealth and the danger of accumulating money without
real responsibility, especially when it comes to interest. However, the

original principle of free giving changed early in the history of Islam into an obligatory and partially formalised duty. The *zakāt* donation has become part of the law that requires every believer to give annually for those charitable purposes that benefit human development or nature. Further, each believer with property is expected to give a percentage of that property's value to benefit others. In this regard, we read a surah in which not only the purpose of these donations is described in more detail, but which in principle contains interesting content regarding an economy of love:

> Love does not consist in turning your faces to the east and
> to the west. It consists in believing in Allah, the Last Day,
> the angels, the Book, and the prophets, and the giving of
> money, although you love it, to relatives, orphans, travellers,
> questioners and the poor, and spending it for the liberation of
> the unselfish, and performing the prayer and paying the tax.
> (Surah 2:177)

The economy of love is about making the entire value chain transparent to guarantee equal opportunities for those involved.

By this we mean the promotion of human development (for example, through education) and thus giving part of the profits to the cultural life. However, this is not charity, but voluntary sharing. If one gives away money, though one loves it, it can benefit humanity in multiple ways. For it is not only the material that is of value to the economy, but the spiritual and the creative, too, as we demonstrated earlier in our approach to the product development of our NatureTex dolls. In the aforementioned surah a reference can also be recognised in our concern for always moving from faith, knowledge or spiritual achievement to action: love does not consist in thinking or praying ('turning your faces to the east and to the west'), but in giving, in active doing.

In Sekem, the profits from the farms flow especially into cultural life: at least 10% goes to educational institutions, research and community development. Ibrahim Abouleish found this approach in the anthroposophical threefold model as well as in the Quran and in Islamic history. He explains the historic flourishing of Islam in Baghdad and on the Iberian Peninsula by the fact that wealthy rulers spent their riches on promoting science, research and culture, and their ministers and rich merchants tried to do the same. The tax on the rich, the *zakāt*, is one of the five pillars of Islam. One of the many surahs on this subject states:

> Believe in Allah and His Messenger and spend from that over
> which He has appointed you stewards. Because for those of
> you who believe and give donations, there is a great reward.
> (Surah 57:7)

This surah makes clear that wealth is not property; rather, human beings are fiduciary stewards of what they have received from nature. In this context, it is also worth mentioning the Islamic *waqf*, an endowment model that emerged during the heyday of Islam in Cordoba. By this model, private property is turned into common property with a charitable purpose, which does not necessarily have to have a religious connection.

Islamic values also help us to better understand the dynamics of give and take:

Who is it that lends a handsome loan to Allah? He will double it to him many times over. And Allah clenches (takes) and spreads (gives). And to Him you will be returned. (Surah 2:245)

Here, the conscious use and the steady flow of products or money is addressed. Moderate, balanced giving and taking is a key idea of the economy of love. It includes an awareness of such things as the origin and impact of production, and the need to share and do things together.

Forty years of an economy in the desert

As of 2022 the valuation of Sekem is around 2 billion Egyptian pounds (around $440 million / £338 million according to the World Bank's calculation). Looking at these figures, we can state with satisfaction that we have managed the last forty years successfully. Today, the Sekem

The company NatureTex processes biodynamically grown organic cotton into textiles.

Group includes companies that cover a wide variety of needs: from agriculture, through raw-material processing and food production, to textile and pharmaceutical manufacturing. In forty years, the vision of the founder has been abundantly realised. The majority of products are sold on the local Egyptian market, while around 30% are exported. The Sekem companies are now well known in the international organic sector as suppliers of high-quality, Demeter-certified goods. Sekem raw materials can be found in the products of many large organic food producers, and Demeter-certified Sekem brand items can be purchased online and increasingly through specialist retailers in Europe. Sekem's NatureTex company produces children's textiles and toys for the international market.

In Egypt the Sekem company Isis Organic has not only been an organic pioneer but also the Egyptian market leader in herbal teas for decades, enabling it to hold its own against multinational corporations. Every pharmacy in the country carries Sekem health teas and herbal medicines from ATOS Pharma.

These successes by no means represent the entire picture of the Sekem vision for the economic dimension. Although the companies want to operate competitively, the focus should always be on people. Therefore, co-workers and producers are supported in many ways, and the profits of the companies benefit the community. In addition, Sekem has now long-standing, trusting partnerships with sustainably oriented financial institutions.

Challenges and opportunities

For millennia, the mainstay of Egypt's economy was agriculture. Since the 1960s, however, it has been in steady decline. From day to day, the population increases and fertile land disappears. Agriculture currently accounts for only 13% of the gross domestic product, and only about 30% of the people are employed in the agricultural sector. Future prospects do not promise any improvement. Low water resources, little fertile land, ongoing soil erosion and the effects of climate change, combined with rapid population growth, increase the pressure to ensure greater food security. Pesticides and artificial fertilisers are destroying the soil and polluting the water, and the air quality is steadily declining due to the burning of waste, industrial activities, and exhaust from cars and other forms of transportation. Egypt's dependence on imported staple foods thus grows steadily. The country is one of the world's largest importers of wheat, for example. Industrial agriculture no longer offers Egypt's roughly seven million small farmers a chance to survive economically. One of the reasons for this is that the value chains are out of balance. In Egypt, the consumer price is often five times higher than what the producer receives. This is despite the fact that globally, and in Egypt especially, we are so existentially dependent on agriculture. A completely new system is needed for the entire agricultural sector to have any future at all.

While industry provides only 17% of all jobs, it contributes nearly 40% of the gross domestic product. But appearances are deceptive. The industrial sector, which seems promising at first glance, is mainly focused on business areas that are not capable of long-term development. Egyptian industry is commodity-based, while globally the industrial sector is already much more focused on knowledge, expertise and technology and will be even more so in the future. Petroleum processing, cement, fertiliser, iron, steel and aluminium production make up a large share of Egypt's economy. These raw materials are finite and non-renewable and thus offer limited prospects for the future. A new approach is urgently needed.

The service sector is Egypt's largest industry, accounting for over half of all jobs. But here, too, there is little focus on services relevant to the future. One reason for this is Egypt's education system, which is rated as one of the worst in the world. There is a severe lack of innovative approaches that would enable the large number of young people in Egypt to meet the various demands of tomorrow's services.

Currently, no alternative exists to the commercial financial sector, although ethical approaches in Islamic cultural areas would offer ideal conditions for this. Banks and investment companies focus on quick profits. The high unemployment rate in Egypt affects mainly women and young people. The resulting poverty and growing income disparities put pressure on the economy and politics. In 2011, the

World Bank classified 25% of Egypt's population as living below the national poverty line. The small but increasingly affluent upper class, which can easily afford to buy sustainable products, prefers instead to reach for the image-boosting offerings of multinational corporations and is characterised by consumption in which prestige takes precedence over quality and sustainability. The preponderant middle class is increasingly imitating this consumer behaviour. In its present form, Egypt's economic life can hardly respond to current and future challenges.

Vision Goal 10: Circular economy

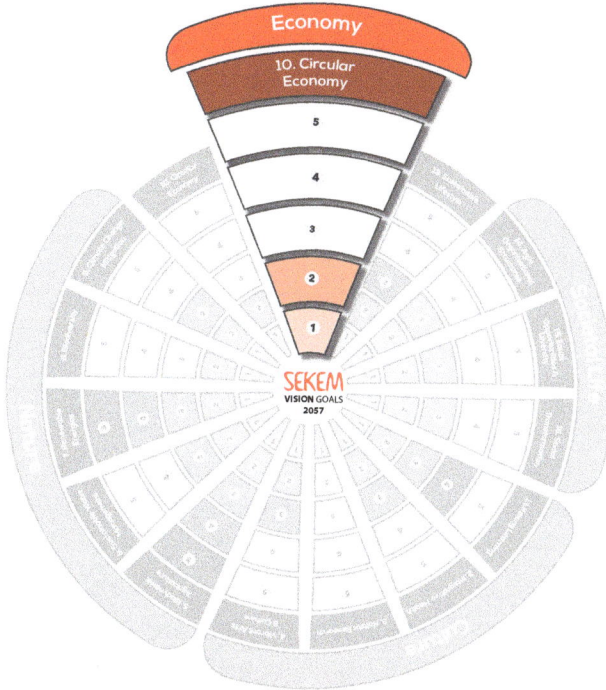

Vision for Egypt 2057
Businesses in the country practise a circular economy. Egypt is a showcase for waste-reduction management.

Progress
For us, circular economics is an inevitable solution for addressing the challenges of increasingly scarce resources and huge waste production. According to current estimates by the World Bank, by 2050, 70% more waste will be produced than today. These masses of waste will be a burden for many generations to come in Egypt. Yet the solution is simple. Indeed, nature shows us how: there is no waste, only raw materials in the wrong place. In a circular economy, waste becomes valuable so that no raw material is lost. Instead, it serves a purpose in a different form. To achieve this, we must include all manufacturing and consumption processes.

Starting with design, we want to develop products using only reusable materials, which can then be used in the manufacture of a new product after the end of one product's life. We also need efficient manufacturing and processing techniques, such as using wastewater or waste heat, and we want to promote longer product life through appropriate design, sharing, repairing or repurposing. Finally, a circular economy is about recycling – that is, recovering the basic materials from used products and reusing them. We have 2 ongoing projects and activities:

- a waste-disposal system for Sekem businesses and community
- the recycling of all organic waster

We would first like to introduce this regenerative economic system ourselves across the board. To this end, all processes are to be examined and adapted to C2C (cradle to cradle) principles. C2C principles describe products that can either be returned to biological cycles as biological nutrients or kept in technical cycles as 'technical nutrients'. C2C certification evaluates five criteria: material health, recyclability, use of renewable energy, responsible use of water and social justice. Heliopolis University incorporates these topics into its curricula and cooperates with the international association Cradle to Cradle, which

has made it its mission to lobby for the rethinking and implementation of circular economies. Together with Cradle to Cradle we are expanding knowledge and competencies around the circular economy to design further application models.

We have been steadily reducing the waste generated by our institutions, so that by 2020 only 12% was non-recyclable (in 2010, the figure was 42%). By 2027, we want to have redesigned all our packaging so that it is 100% compostable and we leave no non-recyclable waste behind.

Inspiration: compost

Compost is the best example of how waste material can bring great benefits in a different context. We use all our biological waste, from agriculture as well as from companies, for composting. To prevent it from ending up in landfills and thus harming the environment by releasing methane gases, we return it to the biological cycle. In combination with the manure from our cows and biodynamic plant preparations, biological waste becomes the ideal fertiliser for our soils. Without compost, we would not be able to green desert soil and keep it fertile in the long term.

Vision Goal 11: The economy of love

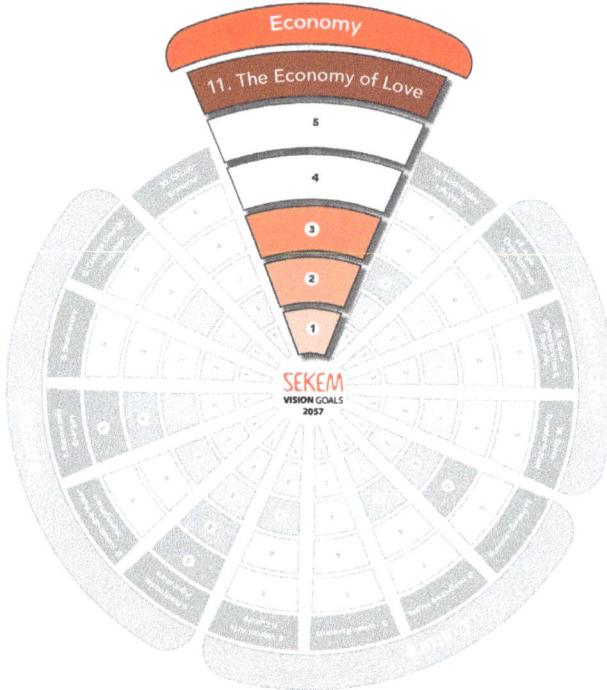

Vision for Egypt 2057

Businesses in Egypt operate according to the principles of the 'economy of love', emphasising transparency and considering true costs.

Progress

As previously mentioned, we began to implement an economy of love decades ago and have implemented this in many areas. However, much remains to be done. By 2057 we expect to incorporate it for all Sekem businesses, and to do so for other economic interests in Egypt. To this end, we have developed standards for the certification of the Economy of Love (EoL) together with local and international inspection bodies. We have 6 ongoing projects and activities, including:

- introduction of EoL (Economy of Love) certification in Egypt and Europe

Demeter or Fairtrade standards are the minimum requirements for EoL certification. This is because the economy of love not only takes into account the farmers or traders who supply the raw materials and packaging for a product, but also the forestry worker who cuts down the tree for the paper from which the packaging is made and the driver who transports the paper. We want to ensure transparency from start to finish, not least because we are convinced that only then can customers make an informed purchasing decision.

We want consumers to be able to answer four basic questions about the product they have chosen, namely:

1. What is the impact of the product and its production on the social environment?
2. What is the impact of the product and its production on the natural environment?
3. What is the impact of the product and its production on the development and potential of people?
4. What is the true price of the product?

Only when enough transparency has been created for these four questions to be answered satisfactorily can care also be taken to ensure that economic activity does not harm the environment or people, but helps them instead.

Including these four aspects – the impact on the social and natural environments and on the individual, thereby arriving at the true cost of a product and the way it is produced – and making them transparent constitutes one of the biggest differences compared to other production standards. The cultural aspect – that is, the question of how a product influences the potential development of both consumers and producers – has so far been given little or no consideration in production. EoL certification is intended to be a multidimensional standard for holistic sustainable development. This means, for example, that producers must always have access to regular cultural activities and educational opportunities.

EoL certification also offers important advantages over organic or Fairtrade certification. After all, water consumption, carbon footprint, or use of renewable energy are equally important for all responsible agriculture, and promoting human potential and community development are of comparable importance to ethical businesses as a fair salary or social security.

A traceability tool makes the entire value chain visible to all. The tool not only provides information about the farmers and processors, but also about transport, the ecological footprint and the actual costs of production (that is, the costs incurred for water consumption, water treatment, air purification, energy or CO_2 emissions – see Vision Goal 5: Agriculture). The EoL standard takes these actual production costs into account for the first time by means of certification.

It is our goal that by 2027 all our own products will be EoL certified and as many as possible of our partner companies that we supply with raw materials. In the long term not only food, but all products and services will be EoL certified. In this way, we want to show that a sustainable economy that benefits the environment and people, rather than exploiting them, can be just as successful and financially rewarding as exploitative models.

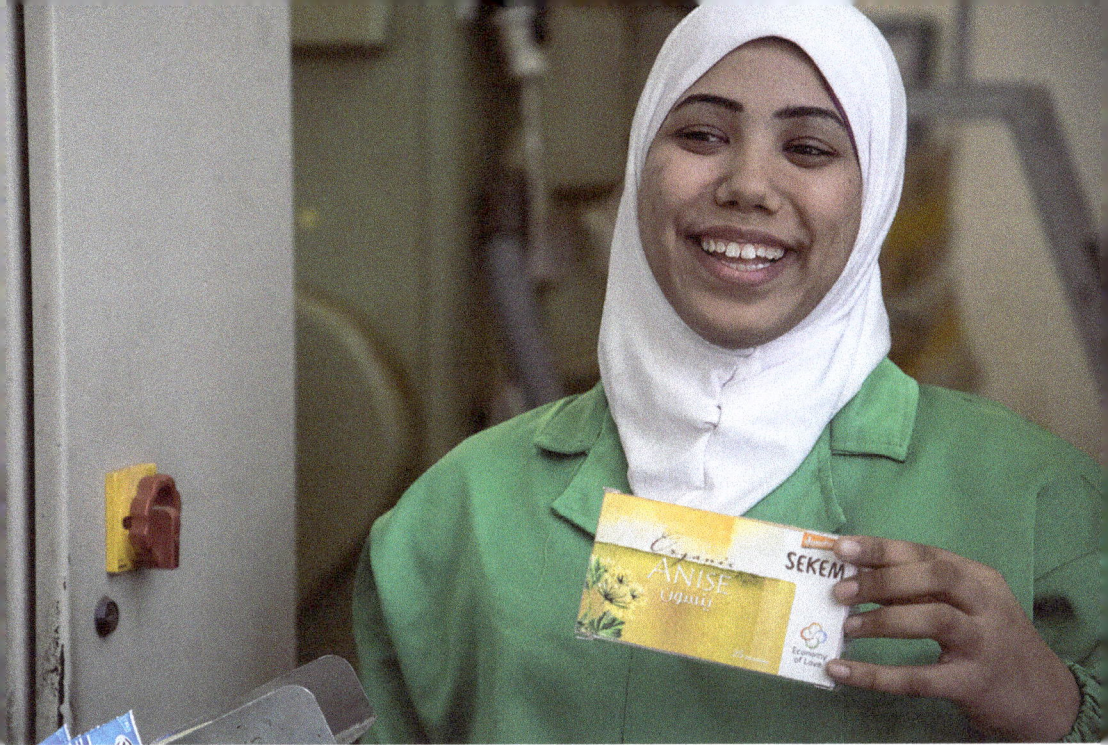

Inspiration: the economy of love is efficient and cost-effective

EoL certification will allow consumers to trace products in a fully transparent way and know the true cost of them. The current pricing system perpetuates a kind of illusion, which is that not all factors are reflected in the price. The EoL standard makes clear that the prices on the shelves do not represent the real value. As an example, consider our anise tea. If we compare the cost of a box of Sekem anise tea with a conventional competitor's product in Egypt, we find the following. The conventional tea costs about 20% less than the same amount of our organic tea (16 Egyptian pounds, or around $0.35 / £0.27, compared to 20 Egyptian pounds, or $0.40 / £0.31), but in production, the organic tea uses 25 litres less water and does not contaminate groundwater. The treatment of polluted water can be calculated at 5 Egyptian pounds ($0.10 / £0.08) per package. In addition, the cultivation of anise seed for one box of anise tea in organic farming sequesters about 75 grams (2.5 ounces) of CO_2, while conventional farming sequesters none whatsoever. If the CO_2 emissions were to be priced, as proposed by the FAO,[6] at 1 Egyptian pound ($0.02 / £0.02), this would make the conventional product an additional 1 pound per box more expensive. Thus, the conventional competitor's anise tea would already be 22 Egyptian pounds ($0.45 / £0.35) and 10% more expensive than our EoL-certified product.

Vision Goal 12: Ethical banking and finance

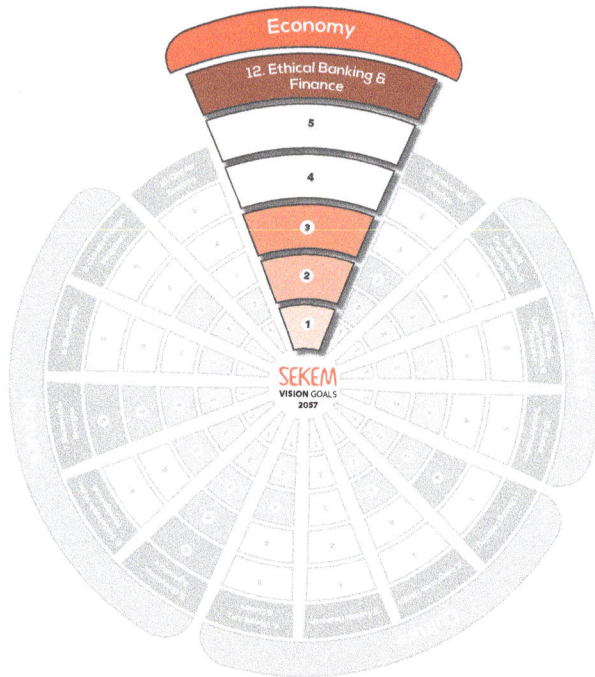

Vision for Egypt 2057
Egypt has implemented ethical banking and finance.
Progress

A truly sustainable economy requires ethical banks that take into account in their operations the environment and the social good. Sekem wants to manage its money in such a way that we know exactly where it is and what happens to it. With the aim of a far-reaching transformation of the Egyptian economy, Sekem works with established financial partners from Europe that have experience providing transparency for customers and consider ethical, environmental and social concerns. If these practices take root in Egypt, banking and finance will come closer to an economy of love. We have 6 ongoing projects and activities, including:

- the complementary payment system, Sekem-Misa
- a microcredit programme for the 13 Sekem villages

First, we established a credit institution in Sekem itself, which started by providing microloans. The microcredit programme manages funds from Sekem companies as well as from donations, and it only grants loans to ecologically or socially sustainable businesses. The resulting lending institution will not remain exclusively available to people living around Sekem Mother Farm, it will also be introduced to other parts of Egypt, such as the emerging community at Wahat Farm (see also Vision Goal 16: Social transformation) and our contract farmers.

Sekem also aims to design and distribute innovative financing models that allow everyone to know exactly where and how the money is being used. A first attempt at this is our 'Greening the Desert' campaign. We take out loans for this purpose and the donors are informed in detail about how this money is being used. They then receive regular updates on the progress of the project. This first exercise in sustainable investing involves financing pivot irrigation, which is needed for greening the desert. Supporters can also help shape the project in three ways:

1.   Through loans, which carry up to 10% interest. The interest is paid out in vouchers that can be redeemed in the Sekem Online Shop for products or for Sekem Travel.
2.   Through the purchase of CO_2 certificates, which can simultaneously reduce one's own ecological footprint.
3.   Through donations to the project.

In this model, both sides benefit, as the lenders receive much higher rates of interest than is currently the case and Sekem can distribute more products to make further sustainable activities feasible.

We are also working on alternative approaches that could make it possible in the longer term to become independent of extreme currency fluctuations. To this end, we have developed the complementary currency Sekem-Misa.

Inspiration: Sekem currency

Sekem-Misa was initially designed as a complementary payment system through which co-workers could earn and pay points when purchasing products in Sekem Stores. One Sekem-Misa is equivalent to the value of one Egyptian pound ($0.02 / £0.01). For every pound spent in a Sekem store, co-workers receive the same value on top in Sekem-Misa for their next purchase. This has increased the demand for our organic products. The complementary currency will be established in the stores of the thirteen villages around Sekem, as well as with our farmers. The Sekem microcredit programme will also use Sekem-Misa with lower interest rates. Farmers and borrowers can use it to buy organic feed or seeds, for example. Thus, the system supports the distribution of sustainable-organic products and a local economic cycle.

Vision Goal 13: Sustainable lifestyle

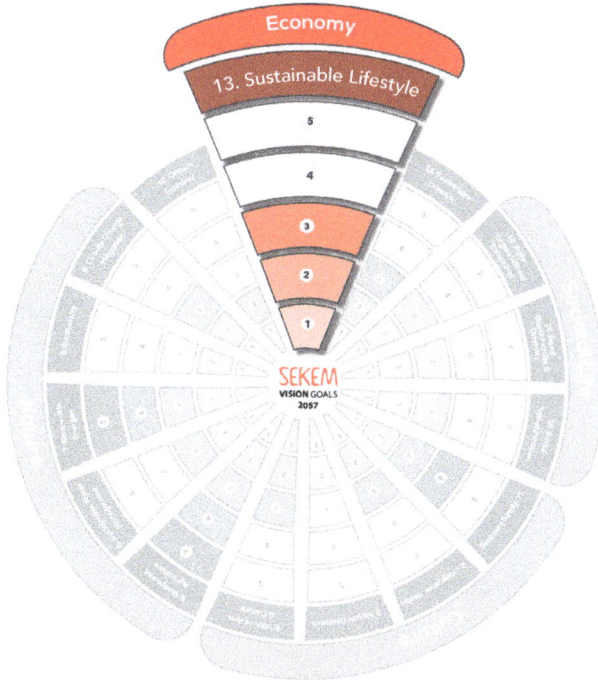

Vision for Egypt 2057

Responsible consumption and a sustainable lifestyle are mainstream. There is a wide range of sustainable products and services for all customer needs and social classes.

Progress

We want to show that it is possible to live a consistently sustainable life, from food to holidays to furnishing your own home. This should involve products and services that bring meaningful benefits to people and support them in their development. We have 4 ongoing projects and activities, including:

- cooperation with Shift Phone
- sustainable architecture in Wahat

Our aim is to raise awareness of the fact that, in principle, there are too many goods on the market that do not meet the requirement of sustainable living. Instead, they clutter our lives and stand in the way of responsible progress. The goal is therefore not to design as many new products as possible, but rather bring to market those that have a meaningful use. The versatile range of existing products can be used in such a way that they provide targeted information and thus promote awareness. For example, Sekem products have become an integral part of the Egyptian market and most Egyptians know them as healthy and high-quality products. But there is little understanding of the importance and benefits of organic, let alone Demeter-quality, products. We have introduced the Demeter-certified Sekem brand and are trying to inform and educate consumers in different ways, such as through communication on packaging, at trade fairs or in the training of sales staff. In the area of packaging, we are working to switch to the most sustainable materials possible.

One important aspect in the design of a sustainable living environment is information technology. In all of our institutions we separate ourselves from the big players in the industry and try to replace technical devices, such as cell phones or computer programs, with offerings from alternative open-source companies.

On our Wahat Farm, we are also building a prototype of a sustainable building that is not a burden on the environment: a single-family house for six people that uses traditional and natural materials paired with innovative technologies. The walls of the house are made mainly from clay, and the interior is made from our own wood. Adiabatic-cooled air, which uses water evaporation to reduce the temperature of warm air, creates a comfortable indoor climate. With the experience gained in the construction of this sustainable home, we hope to create solutions for Egypt's growing building needs.

We do not want to ignore the service sector, either. We have started to organise sustainable travel that allows our friends in Europe to visit Sekem. The purpose is for them to get to know the Sekem initiative, which is helped by excursions into Egypt's rich cultural landscape. This allows us to create awareness for our cause and demonstrate how it works in a local context. Guests will be accommodated in our

Eco Village guesthouse, which is supplied with sustainable products. The climate impact created by a trip can be offset through our CO_2 programme.

For energy, we use green electricity for private households or vehicles (see also Vision Goal 7: Renewable energy). To increase awareness within our community, we carry out exercises that help to define and understand our own consumer behaviour. After all, how often do we realise how big the hurdle is between knowing something and acting on it?

Inspiration: sustainable building construction

The prototype of our sustainable house on Wahat Farm consists of clay (mined from non-agricultural land), wood, linseed oil, glass, steel, natural stone, straw and cotton. Plastics and ceramics are only used for sanitary facilities. Rammed earth has been used as a building material for thousands of years. This form of ecological construction is 'dirt cheap' and therefore affordable for many people. It is resistant to insects, mould and fire; it absorbs sound and is completely non-toxic. Because of its ability to absorb and release moisture from the air, walls maintain a comfortable indoor humidity. As a thermal mass, interiors

can absorb energy during the day when temperatures are higher and radiate it back out at night. Due to its ease of use and pleasant feel, wood is suitable as a building material for all other interior building components and for the roof. Wood can also be CO_2 neutral and is a renewable resource.

The house is protected by a second roof skin, which also serves to promote a permanent exchange of air by removing warm air from the upper rooms and bringing in adiabatically cooled air from below. All energy required in the building is provided by an in-house photovoltaic system. Wastewater, after appropriate treatment, is used to irrigate trees. The consumption of one person is enough for about six trees a day. The trees in turn create locally grown hardwood suitable for furniture production. This allows us to produce sustainable, long-lasting furniture without being dependent on the international timber market. To ensure the efficient and cost-effective use of this precious resource, the design criteria are based on the motto 'less is more'.

8.

Social Life

Beyond ideas of right and wrong lies a field. I'll meet you
there.

Djalal ed Din Rumi

Inspirations for a sustainable society

Our goal at Sekem is to bring new, practical initiatives to the world
that help to heal people and the earth. This requires a place and people
who come together to form a working and learning community; people
who form a social organism in which the spheres of culture, ecology
and economy can thrive. It is in people's encounters during work and
learning that social life takes place, creating a community that can
also be defined and lived in different ways. Thus, there are family
communities, age communities, religious communities, communities
of interest, communities of values and many more. We want to take
into account and include different forms of community, but at the same
time model and promote a community of vision. Ibrahim Abouleish
called it a community of spirit. This community seeks to help people
see themselves as part of a larger world community that is constantly
evolving in a vibrant way. In the words of Ibrahim Abouleish:

> A community in which people of all nations and cultures
> work and learn together in peace – and sound together
> in harmony as in a symphony. A community in which

professions from all walks of life, from all ages, and also from all states of consciousness are represented, especially those that recognise, nurture, and love a super-sensible world – a community that aspires to higher ideals. A living, continually renewing community that maintains its dynamism through the pursuit of a science of the spirit. A community that strives for truth and patience, and that selflessly applies its insights to humanity and the environment.[1]

This vision of community has been shaped by various models, ideas and practical approaches. One of these sources of inspiration is the Islamic *ummah*. The original ideal of the *ummah* encompasses more than just the community of Muslims. It is a community that transcends the framework of a tribe, clan, or nation – a 'community of the middle' as described in the Quran:

> And so, We have made you a community of the middle, that
> you may be witnesses over men, and that the Messenger may
> be witness over you. (Surah 2:143)

The characteristic of the middle is that of a striving for balance in a living process, an equilibrium that is dynamic and not fixed, which is in constant development and includes an understanding and interaction of opposites. The fundamental meaning of the middle existed already in ancient Egyptian culture and was represented by the polar interaction of the divine beings Horus and Seth. Historically this was called *Sema-Tawy*, the 'union of the two countries' of Upper and Lower Egypt, in the middle of which lay the spiritual centre of Heliopolis. By writing this word with the hieroglyphic sign for lung and windpipe, the ancient Egyptians expressed the correspondence of this middle with the human breathing process, which unites the two poles within human beings of thinking and willing.

For us, however, a living, learning 'community of the centre' means not only the uniting of people from diverse backgrounds and the continuous development of those who work in it, but also a connection to the spiritual. Just as work needs willpower and learning needs thought, so social interaction needs 'heart-power' to be alive and to do justice to human beings in their wholeness. Without the involvement of the forces of the heart, the community remains a mere structure of laws that regulate living together, but neither promotes people's potential nor fully exploits their existing potential, which is the prerequisite for true sustainability.

The eighth surah of the Quran describes the necessity of heart forces and spiritual connection for a sustainable and holistic community:

[It was Allah] who makes their hearts sound together. Even
if you had expended all that is on earth, you would not
have harmonised their hearts, but Allah has brought them
together. (Surah 8:63)

This idea is also a foundation of Christianity and is expressed in the
guiding principle of 'love thy neighbour as thyself' (Mark 12:31). There
are also similar formulations in Judaism.

Rudolf Steiner has also repeatedly described the importance of the
feeling level for the communal form of a 'harmony of hearts':

It is the case in every human community that powers flow
to the human being from out of the community; but the
community has to be a real community. It has to be felt,
perceived, experienced.[2]

We try to nourish this connection to spiritual sources, to the heart
forces of feeling and sensing, through spiritual work that makes it
possible to experience this every day. For example, in the different areas
of the Sekem initiative, each day begins with a morning circle that
brings the members of these smaller communities together and makes
it possible for them to feel that they belong to a larger community with
shared values. Standing next to each other – that is, on the same level
and despite whatever differences may exist between them – a centre
is created both figuratively and at a sensory level. Each individual
experiences themselves as part of a community of equals and feels
called upon in their actions to take those around them into their
consciousness for the good of the whole. Inspiring words are spoken
together while people join hands, and thus the heart forces and feelings
are addressed.[3]

This ritual also combines different impulses from different cultures.
The Islamic greeting 'Peace be with you, and God's blessing and
mercy!' is spoken together, and the dates of the Islamic, Gregorian and
planetary calendars are mentioned. This is followed by a verse from
Rudolf Steiner that addresses the wholeness of social togetherness with
its different qualities:

To wonder at beauty,
Stand guard over truth,
Look up to the noble,
Decide for the good:
Leads human beings on their journeys
To goals for their live,
To right in their doing
To peace in their feelings
To light in their thoughts
And teaches them trust,
In the guidance of God
In all that there is:
In the world-wide All,
In the soul's deep soil.[4]

In addition to their work in Sekem's various operations and institutions, all co-workers regularly take part in meetings and activities that appeal to their artistic interests, further their professional knowledge or deal with social issues. Ibrahim Abouleish believed that constant spiritual endeavour provides a viable foundation for all that Sekem aspires to become.

Within the framework outlined here, awareness is raised of values that are necessary to create a just community in which individual potential is nurtured, from human rights to equal opportunities for women and men, to leadership structures. In this way, we also aim to promote an awareness that what is at stake here is a togetherness that transcends the interests of individuals, family ties or faith communities. Rather, we want to convey a vision of our community contributing to a world community that unites all people, regardless of faith, ethnic origin or social status. In this way, diversity can at the same time be a unity, just as nature shows us and as Goethe poetically describes in *Faust*: 'How everything weaves itself into the whole, the one works and lives in the other.'[5]

The Quran also celebrates differences within human communities. The differences show the diversity of life in which the different stages of human development are expressed.

For each of you, We have set a direction and a path. And if
Allah had willed, He would have made you one community.
But He wants to test you in what He has given you. So,
hasten on to the good things. To Allah you will all return,
then He will make known to you what you have disagreed
about. (Surah 5:48)

For us, such a community of unity in diversity also means that the
potential of individuals can multiply and increase many times over
through mutual exchange. A community grows when the individual
learns to think in terms of the whole; this requires at the same time that
the individual is heard and taken seriously. This is also an idea that can
be found in anthroposophy:

A healthy social life is found only when in the mirror of each
soul the whole community finds its reflection, and when in
the whole community the virtue of each one is living.[6]

The Quran teaches openness to different communities and
individuals, even more to the freedom of the individual. In Sekem,
this has always been an indication for us that in the Islamic faith
other forms of religion are respected and have their justification. An
indication of this is given in Surah 10:99:

If your Lord willed, those who are on earth would all become
believers together. Is it you who can force people to become
believers?

And even more specifically, Surah 2:256 states:

There is no compulsion in religion. The right change is now
clearly distinguished from the wrong way.

In the esoteric currents of Islam, which were of great importance for
Ibrahim Abouleish, religion is understood as a state of mind, not bound
to any denomination. The Islamic mystic Rumi wrote:

> Not Christian or Jew or Muslim, not Hindu, Buddhist, Sufi
> or zen ... My place is placeless, a trace of the traceless ... I
> belong to the beloved, have seen the two worlds as one.[7]

And it was Rumi who said more than seven hundred years ago: 'The lovers of God have no religion but God alone.'[8]

For the sustainable community that Sekem aspires to be so that it can shape larger society in constructive ways, the essence and values of social cohesion are primary. This involves questions of law, equality, peace and freedom for each individual, but also for society as a whole, including its political structures. Such basic values are described in the Quran in Surah 3:104:

> From among you shall arise a community of believers
> who call for good, enjoin what is right and forbid what is
> reprehensible. These are the highly fortunate.

Or in Surah 7:181:

> And among those We have created is a community of people
> who guide by the truth and act justly according to it.

In addition to these Islamic principles, the anthroposophical concept of the threefold social organism also serves as a guide for shaping the community. According to Rudolf Steiner, the threefold model aims to transform the structure of society from one of centralised control to one consisting of three autonomous spheres: the cultural or spiritual life (*freedom* in culture, education and science), the life of rights (*equality* in laws and social rules) and economic life (*fraternity* in production, trade and consumption).[9] The three spheres are on an equal footing, but the cultural sphere provides values and ideals for all spheres. Each sphere is autonomous, but they cooperate with each other. At Sekem these spheres constantly interact but without surrendering their autonomy: the cultural institutions could not exist without economic activity, and the economic enterprises could not exist without educational foundations and intellectual creativity.

A verse from the Quran illustrates that this tripartite division of society did not originate with anthroposophy:

> The believing men and the believing women are one another's friend. They enjoin the good and forbid the reprehensible, perform the prayer, pay the tribute for the poor, and obey Allah and his Messenger. (Surah 9:71)

Introduced by reference to the equality of people (note the mentioning of the equality of women and men), the surah follows with references to legal life ('enjoin the good and forbid the reprehensible'), spiritual life ('perform prayer') and economic life ('pay the poor tribute').

Equality of opportunity, which we try to practise within the community, is also an Islamic ideal, because the Quran says, 'He is the One who made you deputies on earth' (Surah 35:39). The word *khalifa*, used here for 'deputy', is not used in the Quran in connection with a privilege or in connection with a head of state (as it is often understood when referring to the caliphs, the rulers after the Prophet Muhammad). Rather it means the special position of all human beings regarded equally, in that they are God's representatives on earth, responsible for themselves and for creation (see Surah 2:30). Thus, the term *khalifa* points to the unity and brotherhood of all humanity.[10]

What has been said demonstrates that the communal ideal of Sekem occurs just as much in the Islamic cultural sphere as it does in the Christian one, because it is valid throughout the ages for all human society. Today's humanity, however, has moved far away from this basic condition. It was because of this that Ibrahim Abouleish went into the desert to form a new community from nothing, free from prevailing social norms, to change existing social structures. His goal was not a utopian enclave, but a seed from which society can be renewed.

Forty years of community development in the desert

As we have already mentioned, at the start of every day, people gather across the various Sekem initiatives and stand in circles to reflect on their shared commitment. Around two thousand people of all ages, cultures, religions and professions stand like this every morning. They represent only a part of the Sekem community, however, because in addition to the co-workers there are people all over the world who are directly or indirectly connected with Sekem – starting with the international guests of the Sekem Guesthouse, the members of the four European Sekem Friends Associations and the consumers of Sekem products worldwide. Then there are the thousands of people who are in direct contact with Sekem: local farmers, our industrial and financial partners, and people from the surrounding villages who visit Sekem's medical centre. There are a growing number of students from Heliopolis University, as well as the university co-workers and companies on the Sekem Mother Farm, who are active there on a daily basis.

The group most deeply involved in community development is the Sekem Future Council, which has been in place for several years, although this does not mean it is of greater importance than any of the other circles. The Future Council has an awareness of all the other circles and works to ensure that the Sekem vision is realised in its entirety without any one area or personal interests dominating the others. All these areas are interwoven and of equal importance for the further development of the Sekem community and its impulse.

The connections with diverse people and the multifaceted partnerships across the globe that Sekem has forged in forty-five years continue to be important for the continuation of community life. Especially during challenging times – and there have been many – we have been able to rely on loyal partnerships and trustworthy support. This has helped us to overcome various crises without jeopardising our vision for a sustainable community. Without this network, Sekem could not exist.

Challenges and opportunities

While a holistically oriented community for a sustainable future has been developing in Sekem since 1977, society in general in Egypt seems to be increasingly in crisis. In the mid-twentieth century it was still characterised by a relatively open and free life with good prospects for progress. But problematic times followed and with them a widespread deterioration of infrastructure, the education system, food supply and much more. The revolution of 2011, which sparked great hope, has led to little change. Although the country's economic situation has stabilised relatively well compared to the enormous challenges of recent years, the mood among the population is one of resignation.

Population growth remains the greatest challenge facing Egyptian society. Studies estimate the population will reach 150 million by 2050 (up from about 100 million in 2020, and about 20 million in the late 1950s). As in many other parts of the world, Egypt expects a growing rural exodus over the coming decades. This will be accompanied by unemployment, poverty and problems with the food supply, among other things. This could lead to increased migration to Europe. The Egyptian government wants to counteract this with various projects. There is currently the 'Two Is Enough' campaign, which aims to curtail the birth rate, and a plan to reclaim around 650,000 hectares (1.6 million acres) of desert for conventional food production by 2030. But these so-called mega-projects have not yet been as successful as hoped. The reasons for this relate to the persistently strong top-down structures in society that lack innovation and result in fewer opportunities for people to participate.

Other hurdles include the lack of diversity and equality. Women continue to have fewer opportunities than men and only 20% are employed. As a result, Egyptian society foregoes much of its existing potential in terms of skills. Minorities, such as people with disabilities, are also barely integrated into social life. And even for young people, who represent the future and make up over 60% of the population, there is a lack of opportunity for them to develop their potential and use their skills.

Organisations, companies and state institutions are largely stuck in hierarchical and bureaucratic structures and react far too slowly to the needs of the times. A paucity of contemporary approaches to organisational development in companies and institutions results in very low efficiency, productivity and agility. Existing structures create a vicious cycle of low pay, overstaffing and low performance.

The policies of President Gamal Abdel Nasser in the 1950s and 1960s brought more culture and social security, but they also contributed to the population's loss of its own sense of responsibility for shaping the country. Over the course of subsequent presidencies, the conviction prevailed that the government would take care of everything, and awareness of one's own autonomy disappeared.

Egypt therefore needs a social transformation that enables people to regain their sense of responsibility for taking initiative to shape their country through social security and sustainable structures.

Vision Goal 14: Agile organisational structures

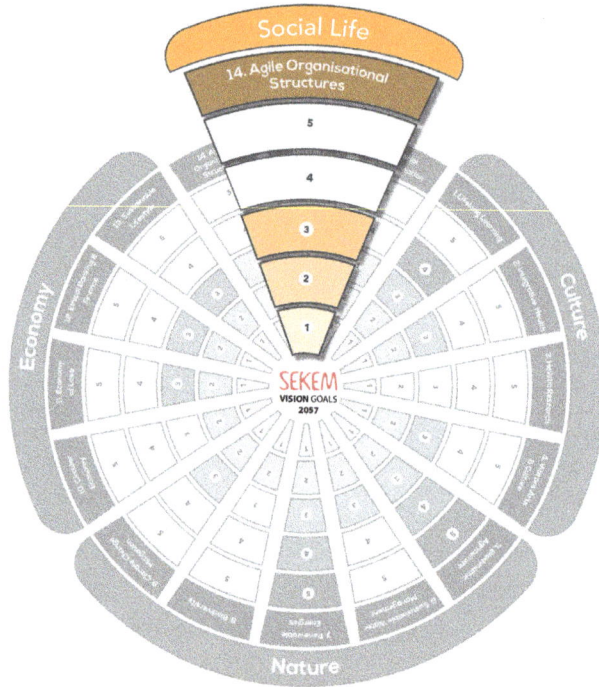

Vision for Egypt 2057

Across various organisations in Egypt there are vibrant and agile organisational management practices that are tailored to people's development and state of consciousness.

Progress

All of us are engaged in a transformation process with the organisation and administration of Sekem. This means that to ensure the sustainable health of the Sekem organism we need to examine and update structures shaped during Sekem's founding period. In this way, we are learning what it means to create living organisational structures that not only guarantee stability but are resilient and promote development at the same time.

For a sustainable society, we need modern, agile structures that take into account people's (and communities') different states of consciousness. We have 2 ongoing projects and activities, including:

* the Sekem Sophia programme

We started the Sekem Sophia programme in 2018 to address the issue of organisational development from different perspectives and through a wide variety of activities. The programme has helped us to restructure the organisation of Sekem in a more resilient way, and it aims to create approaches for other institutions to follow. The programme asks the question: how do we want to shape our living and working together, and in what way are we able to do this? It is based on various theories and models, including the organisational development theories of Friedrich Glasl and the work of Frederic Laloux and Brian Robertson.[11] Of particular importance in the Sekem Sophia approach is the acknowledgement that individual members are at different stages in their development and each possess different qualities of consciousness, and that this is also true of the culture of organisations as a whole (see page 152 and 'The development of consciousness').

In NatureTex and ATOS Pharma, for example, we have started to apply concepts from the Sekem Sophia programme. Responsibilities are defined in terms of roles, which can be fulfilled in a variety of ways in a wide variety of circles. We also aim to reduce hierarchies, strengthen personal responsibility and become more agile, which will enable us to become more resilient. In the recruitment, training and development of personnel we use the Spiral Dynamics model developed by Don Beck and Christopher Cowan, which describes how people, organisations and societies evolve. It is based on the idea that people progress through different stages of awareness or consciousness. This allows for a much more individualised approach to different personality types, which increases co-worker performance and satisfaction, simplifies administration and organisation, and at the same time increases efficiency and productivity.

Another application of the Sekem Sophia programme relates to the handling of property. Who actually owns Sekem? The Sekem initiative

and all its related institutions are no longer personally owned but held in trust. A majority of the original owners of Sekem (61.5%) transferred their shares to a charitable foundation governed by members of the Sekem Future Council, who have 100% voting rights. This safeguards the concerns of the Sekem vision and its implementation.

To develop, implement and apply an official Sekem Sophia programme, a centre of excellence is being planned at Heliopolis University, as well as a consulting firm within the Sekem Group of companies.

Inspiration: making meetings more effective

NatureTex was the first company to introduce new meeting formats and systems. Initially, this was not very well received. Meetings became more structured and contributions were no longer impromptu. Topics could not be introduced indiscriminately and instead had to be submitted in advance via a digital platform. This was a challenge and aroused resistance. Over time, however, people became more open to it, and now, after about two years, the meetings are functioning much more effectively. The so-called 'check-in', acknowledging participants with a few personal words, is now appreciated and carried out routinely. The initial fears of being caught up in an impersonal system have since given way to an efficient and at the same time personal basis for meetings.

Vision Goal 15: Diversity and equal opportunities

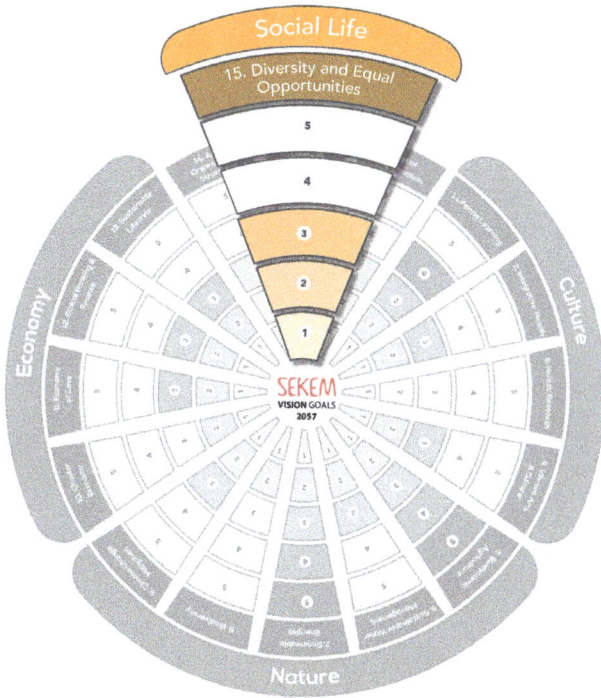

Vision for Egypt 2057
Egypt celebrates diversity in society and ensures equal opportunities for all, regardless of age, nationality, religion or gender.
Progress
Without the opportunity for all people to freely develop their potential in society and contribute to the whole, there is no sustainable development. Therefore, we want to ensure at all levels that diversity in our community increases and that all members have the same opportunities. We have 4 ongoing projects and activities, including: • 'Harassment-free Campus' initiative at Heliopolis University • Increasing the proportion of women in employment

In the Sekem schools, from the very beginning, children of different faiths learn all subjects together. The minority of Christian children have their own prayer room with an altar and receive Christian religious instruction, which is not common practice in Egypt.

An important step towards equal opportunities for women and men is our 'Gender Strategy for a Balanced Society', in which we outline how we intend to support women in exercising their rights. For example, we carry out educational work, help women balance family and career, and commit ourselves through a women's quota to promoting female staff and appreciating their qualities in a wide range of departments.

The integration of people with disabilities in our companies has always been part of the norm, and our training centres for physically or mentally impaired people offer a model for Egypt. We guarantee all students in special education a job in one of our companies with a salary that is equal to that of their non-disabled colleagues. People with disabilities are highly valued in Sekem and make an important contribution to community life with their skills.

Sekem has a gender strategy to ensure equal opportunities.

In the near future, we are planning a multi-generational house and a social model that does not exclude older people from community life, but continues to give them tasks and a role, integrating them into society and valuing their contributions.

Inspiration: advancement of women

Our 'Gender Strategy for a Balanced Society' helps us to support women in exercising their rights and developing their potential. In various ways we help women to stay in work. For example, we have set up a baby group and childcare facilities, women benefit from flexible working so that they can breastfeed their babies during work hours, and men can apply for special leave when they become fathers. There are talks with experts on topics such as sexuality, female circumcision (which is still practised in Egypt, especially in rural areas) and career planning.

We celebrate International Day of the Girl every year to encourage girls to choose professions that have been traditionally associated with men and, conversely, to arouse interest among boys for female-dominated professions. As a result, the number of female students at the Sekem Vocational Training Centre studying to be carpenters, mechanics and electricians, to give a few examples, has increased significantly within just a few years. We also support women in the thirteen villages around the Sekem Mother Farm in developing their own business models.

Vision Goal 16: Social transformation

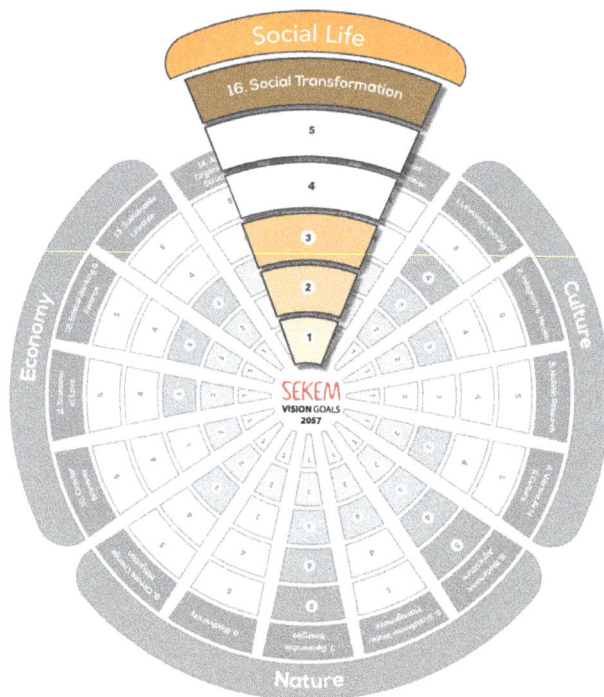

Vision for Egypt 2057

Social transformation in Egypt has led to sustainable rural development and regenerative cities, empowering people to take responsibility for shaping the country's future.

Progress

It is obvious to us and many others that Egypt needs a comprehensive social transformation. At Sekem we work with different approaches to address the major challenges we face. We want to create models that enable the emergence of new structures and developments in the desert. To achieve social transformation, we consider three levels relevant:

- the individual
- the organisational
- the community

The promotion of individual development is implemented at Sekem through the Core Program, that is via artistic and cultural activities. Organisational progress is addressed through the Sekem Sophia programme, and community development is shaped by the Sekem Future Council. In the coming years, the Future Council will be established as an entity that can ensure that the concerns of the Sekem vision are always incorporated into community activities, or that they are chosen to reflect and promote this vision. In the spirit of the Future Council, there will be a vision group for each area that will undertake, on a smaller scale, the tasks that the Future Council has for the overall structure.

We want to promote sustainable community development in Egypt's cities, countryside and desert. We are practising the transformation of pre-existing structures in rural areas through our '13 Villages' project. Here, Sekem cooperates with the surrounding villages in the four main areas of culture, ecology, economy and social development. This starts with training courses on topics such as organic farming, hygiene or setting up a business, and continues with the granting of microcredits. It ends with actual contracts and cooperation in areas such as domestic work, medical care and waste processing. The '13 Villages' project is accompanied by research work carried out by students from Heliopolis University.

Our farm in Wahat El-Bahareyya in the western desert, about

400 kilometres (250 miles) south of Cairo, serves as a testing ground for community development in which there are no pre-established structures. Some years ago, Sekem acquired over 1,000 hectares (2,470 acres) of desert land in Wahat, but relatively few hectares had been cultivated with date palms and jojoba plants. In 2019, we started the 'Greening the Desert' campaign, which has since been used to reclaim large areas of desert and build community structures. In line with the Sekem vision, the desert is sustainably 'greened', jobs are created and people are given the opportunity to develop their potential. Each of these measures allows us to promote the values of transparency, trust and sustainability.

At the outset, the aim is to create fertile soil as a basis for further development, and for that, investments are needed in the desert. In the first phase, we developed 63 hectares (155 acres), which required three pivot-irrigation systems. We financed these through an innovative crowdfunding campaign (see also Vision Goal 12: Ethical banking and finance). Within a very short time, we were able to develop the first 63 hectares of sandy soil. Since then, irrigation has been extended to 233 hectares (575 acres) with eleven pivot systems, all of them completely solar-powered. However, solar power is only available during the daytime, while irrigation is more efficient at night. Therefore, we have started to build reservoirs higher up on nearby hills so that irrigation through gravity can be applied at night.

The conscious use of water in the middle of the desert is particularly important here. The Wahat farm lies above the Nubian Sandstone Aquifer, one of the largest underground water reservoirs in the world. It contains around 1,500 times as much water as flows through the Nile in Egypt each year. Constant research helps us to use the fossil groundwater as sparingly as possible to ensure that the water table does not drop too quickly, and that water extraction is justified by its positive impact on people and the ecosystem.

In addition to agricultural activities, a cultural programme has been developed: a Core Program for co-workers, a theatre and recently a school. Heliopolis University has established an extension on the desert farm, through which students conduct applied field research. More educational facilities and medical care will follow.

The 'Greening the Desert' project on the Wahat farm offers a particularly exciting adventure for us, because every day it shows, in very concrete ways, how in this enormously challenging environment we can bring something new and creative to life.

Inspiration: the social transformation of Sekem's thirteen villages

In the thirteen villages around the Sekem Mother Farm we are testing how our vision goals can be integrated into, or transform, already existing structures. For this purpose, we have developed specific activities for each goal such as: integration of learning methods to develop potential in eight schools, regular cultural events without entrance fees, supply of renewable energy, sustainable waste management, microcredits for sustainable business models, mobile health convoys, which include education and care services, the offer of organic food in supermarkets and much more. The project covers an area of 45,000 square kilometres (17,400 square miles) and reaches around 30,000 people. The goal is that the thirteen villages will soon have reached a standard of living equivalent to that of the Sekem community in all spheres of life.

9.

Our Vision for Egypt 2057 in the Context of Current Events

Although our vision of Egypt 2057 may sound far-fetched, it has been intensively thought through and is based on scientific and sociological research, as well as on years of experience and spiritual-philosophical examination of Egypt and the world. We have not simply created a utopian image of the future, but a vision based on clear awareness of the times in which we currently find ourselves. To do this, we initially asked ourselves what the future means in the first place and what it takes to create the future. We also reflected on the current conditions in which we live while working to fulfil our vision of Sekem in Egypt and the world.

Forging the future out of initiative and expectation

With his vision of Sekem, Ibrahim Abouleish outlined a future that is neither a culmination of what has already been experienced, nor an exuberant utopia torn out of any real-world context. Rather, he painted a picture that is clarified by an awareness of two qualities of the future: one that is 'becoming' and one that is 'coming'. Life is shaped by two streams, one that comes from the past and rushes towards the future, and the other that flows from the future towards the past.[1] The stream that flows from the past to the future we can call *futurum*. It can be understood as a space into which we project our plans that have grown out of our experiences. By contrast, *adventus* represents the future that we cannot foresee but which is coming our way. It is a space filled with great expectations. *Futurum* is characterised by the initiative to do something, by activity: it shapes an empty space. *Adventus* is characterised by serenity, an active waiting and a readiness: it comes to us through destiny.

Ibrahim Abouleish knew how to work with these two streams to create the right momentum. This ability to recognise and act according to the two aspects of the future – to be able to shape a situation through initiative while awaiting whatever was to come with serenity – and then to act with loving devotion, were of decisive importance in the realisation of the Sekem vision. We have endeavoured to work with this understanding of the future in formulating the Sekem Vision Goals and our vision of Egypt for 2057. We have set clear goals and act accordingly, but at the same time we remain open to the unknown and whatever is to come.

Another helpful guide for shaping the future, this time in regard to changing organisational structures, is Otto Scharmer's 'Theory U' model. This management theory aims to help managers reform unhelpful organisational structures and break out of unproductive patterns of behaviour to become more efficient in their decision making. Using this guide, we took the following steps. We looked at the past and current realities, we explored other perspectives and engaged in dialogue, then we tried to free ourselves from everything that had been shaped by the past. We then experienced what Scharmer calls 'presencing' in a completely objective way. For us, it was a matter of reflecting on who we are as a Sekem community, or what the Sekem initiative actually is. We reconnected with the sources and tried to recognise the tasks coming towards us from the future. From this, our vision was able to take shape. Now we are at the point of realising the vision, testing it and developing prototypes of practical action that can then become mainstream.

The first phase of the 'U' in Otto Scharmer's model, the path down to 'presencing', deals with letting go of the past – that is, it is concerned with the *futurum* perspective. The subsequent ascending phase illustrates the principle of active waiting for what comes out of the future and thus the *adventus* aspect of the future. This book can also be seen as a kind of written documentation of the Theory U process, as we have gone through it in Sekem and want to continue going through always anew.

The development of consciousness

The questions about the future, its design and implementation, are important steps towards the realisation of our vision. But they would be meaningless without people. Individual development has always been at the centre of everything we do. Thus, the original Sekem vision was influenced, among other things, by Rudolf Steiner's understanding of the development of consciousness, which we received primarily through the writings of the Egyptologist Frank Teichmann. But there are also other thinkers who have put forward very similar theories about the development of human consciousness. One of them, Jean Gebser, was one of the first consciousness researchers to describe different structures of consciousness.[2] He posited that the next stage of human development will bring about an awareness of the unity of human beings with nature and the universe without losing or eclipsing the previous stages of consciousness. He called this new stage of consciousness the integral stage. Other integral theories, such as the Spiral Dynamics approach, similarly describe the development of human worldview levels. The Sekem community always tries to take these integral approaches to the development of consciousness into account in everyday life. They have also guided us in the elaboration of our vision of the future. We will therefore look a little more closely at two of these models.

In his books and lecture cycles, Rudolf Steiner refers to three levels of soul development: the sentient soul, the intellectual or mind soul, and the consciousness soul. The sentient soul can be characterised primarily by an orientation towards feelings and by a strong sense of belonging to a larger social group. The intellectual or mind soul, on the other hand, focuses on the mind over all things emotional and on one's own self, whereby actions are guided by one's own advantage. The consciousness soul unites the previous stages; it can reflect on them and has a vision of life in its wholeness. With consciousness at this level one can begin to develop a deeper understanding of freedom.[3]

Spiral Dynamics, developed by Don Beck and Christopher Cowan based on the work of psychology professor Clare W. Graves, goes into greater detail. It describes how different worldviews emerge through the interaction of life conditions and our individual mental capacities. This model of evolutionary development is represented by a spiral rising through nine colour-coded levels, each one a different value system. The lower levels of this spiral of values are concerned with matters of survival, magical spirits and the organisation of people into tribes in which the individual has little awareness of themselves as a distinct, differentiated being. This is followed by states in which a self begins to emerge, but which is still subject to rules and authority. The self then separates from the 'we' mentality and begins to strive individually for truth and meaning. Rational scientific achievements dominate, and the earth's resources are exploited for one's own goals until, at a later level, the sense of connectedness returns. This time, however, the individual chooses their own community, and compassion replaces greed, dogma and cold rationality. Finally, the community and the individuals within it achieve integration, and flexibility and spontaneity emerge. Complex, holistic systems form a living, conscious order in the later levels of consciousness.

In both the Spiral Dynamics model and Rudolf Steiner's understanding of consciousness and soul development, it is assumed that it is not possible to skip levels and that living through each level is crucial to the health of the entire developmental spiral.

Every human being passes through different states of consciousness during their life and, at the same time, can express different qualities of consciousness in different situations. But even in entire societies or cultural spheres, certain levels of development may dominate. For example, Egyptian society is characterised by centuries of dominant rule. The majority of people live in familial, community structures in the countryside, but also increasingly in the cities. The population is permeated by a consciousness that assigns decision-making power to a single ruler and the knowledge of what is best for the majority. As a result, a worldview predominates in Egypt in which the collective comes before the individual. This is also evident in the practice of faith. Here, traditions and the dictates of religious authorities are paramount. Individual reflection is hardly widespread. With the Egyptian revolution in 2011, however, it has become clear that a new consciousness is slowly emerging. More and more people are breaking away from such obedience structures and are beginning to question the way things are.

This next level of consciousness for Egypt, most integral thinkers agree, is the one that exists for most of the Western world. Increased rational thinking, strategic efficiency, performance orientation or objective science characterise this worldview. The individual person moves into the foreground and displaces the consciousness concerning the well-being of the community. On the one hand, this creates room for individual development, but on the other, egoism.

Within the Sekem community, different structures of consciousness with different views and interests come together. There are various groups each with their own tasks to fulfil, whether that is producing and selling as many goods as possible, or promoting education or cultural events. And within these groups each individual lives with a certain prevailing consciousness. Thus, there are individuals who are more characterised by sentient-soul thinking, which manifests itself in a pronounced faith or a strong need for guidance and rules, and then there are those with a pronounced mind-soul consciousness, which manifests itself in the fact that they need evidence for everything or expect a direct personal benefit.

These different perspectives result in correspondingly different interpretations of certain developments. Thus, some people will see challenges and difficulties where others see opportunities for positive development. One example is the population growth in Egypt. For some, this represents one of the greatest challenges of the future; for others, it holds enormous opportunities for economic growth. Still others do not even think of family planning, let alone question the number of children, because children are always and everywhere gifts from God.

When the Sekem community comes together every Thursday in the big end-of-week circle, there are several hundred people standing next to each other, people with very different perspectives and levels of consciousness. They all see a community, but each participant represents their own particular worldview. One or the other sees the people next to them as sisters and brothers, as children of God, and feels as if they are part of one big family. Others probably feel more that they are standing next to colleagues with whom they do good work. And then there will be people standing there who see an overarching common vision on the basis of which we all come together.

Sekem has made it its aim to set goals and impart inspiration, to create an environment in which people can develop according to their individual possibilities and where different states of consciousness are considered. This should always be as much about one's own development as well that of others, so that at some point the transition to a holistic, integral consciousness can be achieved. Our vision of Egypt in 2057 was designed against this background and aims to address different levels of consciousness in practical, everyday ways. This claim may sound logical and comprehensible, but it is the greatest challenge. Political, social or economic crises arise in part because there is a lack of personal understanding for the different levels of consciousness. A new awareness of these levels and the opportunities for growth could contribute to healing many global problems.

Conclusion

It is not enough to know, one must also apply; it is not enough to want, one must also act.

Johann Wolfgang von Goethe

In this book we have laid out Sekem's developmental process, from evolving a vision and setting goals to their eventual implementation – that is, the journey from inspirations to practical models that can contribute to real transformation. This process was started by Ibrahim Abouleish, who created the Sekem vision, and we have spent forty years learning and pioneering together with him. In our vision of Egypt 2057 and our intermediate goals for Sekem 2027, we now want to achieve the widespread acceptance that will support systemic change.

We also want to state our fundamental intention to work with a heartfelt connection to our spiritual sources of inspiration. This is because we believe that we can only tackle and change the problems of today in a sustainable way if we also engage and connect with them on a spiritual level. In doing so, we must always consider our own consciousness as well as that of the other person and the circumstances in which we encounter each other. We want to engage fully with heart, head and hands in the challenge as well as in the solution.

In our annual reports, in our regular news (the *Sekem News* sent worldwide, available to order at www.sekem.com), on our website, in webinars and personal exchanges, we keep our friends and partners informed as to the progress towards realising our vision of Egypt 2057. However, for this vision of the future to become a reality, many more hearts, minds and hands are needed. We welcome anyone and everyone who wants to support us. Only together can we shape and welcome a future that fills us all with joy.

Symphony

who would not like
in every moment
to hear
the symphony of life?

who would not
be a note in it
maybe even a player
in the orchestra?

have you forgotten
that we all
had a letter
sent to us
at birth
directly from the great
conductor
inviting us
to bring
all our gifts
as an instrument

but only a few
have ever read the letter
not trusting
that the invitation
was actually meant for
us

so why not today?

look
the envelope smiles in
your heart
longingly waiting
for the hour
when you will open it
and fully
accept
the invitation.

Dedicated to Ibrahim Abouleish
Channelled by Alexander Schieffer

The Sekem Future Council Members (2025)

Full members

Helmy Abouleish is co-founder and chief executive of the Sekem Initiative.

Konstanze Abouleish was the managing director of NatureTex, Sekem's organic textile company. She is involved in music and cultural projects.

Maximilian Abouleish-Boes is responsible for monitoring the sustainability of Sekem's activities.

Mona Abouleish-Lenzen is a eurythmist and is involved in the development of pedagogy and cultural programs at Heliopolis University.

Rembert Biemond is an entrepreneur, professional director and member of the Sekem Board of Directors. He lives in Sweden and is involved in the funding and organisational development of Sekem.

Rafik Costandi is a teacher at Sekem School and supports the educational institutions administration.

Martina Dinkel is a eurythmist and organises the Sekem Core Program and Space of Culture activities.

Yvonne Floride is a visual artist and educator and leads the visual arts programme, especially within the Core Program at Heliopolis University.

Christoph Floride is the manager director of Lotus, the company that processes herbs and spices from Sekem's contracted farmers.

Regina Hanel manages the Sekem kindergarten and assists with international communications and administration at Heliopolis University.

Angela Hofmann is a biodynamic farmer and the deputy director of the Egyptian Biodynamic Association. She is responsible for animal welfare at Sekem.

Associate members

Christine Arlt has worked in the public relations department at Sekem as well as in international project management. She is now based in Germany where she heads the German Sekem Friends Association.

Thomas Fischer is secretary of the Sekem Board and works on expanding the distribution network for Sekem products in Europe. He lived in Sekem for many years and is now based in Germany.

Justus Harm coordinates the Egyptian Biodynamic Association, oversees the economy of love certification program and takes photographs and produces promotional film clips for Sekem.

Samuel Knaus helped develop the complementary currency, Sekem-Misa, and oversees Lotus customer service. He also works as a photographer.

Nana Woo helps to manage the Core Program for students and staff.

Honorary members

Gudrun Abouleish wife of the late Ibrahim Abouleish and a co-founder of Sekem. She has been involved in the development of a wide range of areas at Sekem and today is responsible for the administration of the company and guest operations.

Inge Marienfeld helped to build Sekem's Medical Center and is now involved in the doll workshop at NatureTex.

Dieter Marienfeld takes care of the administration and maintenance of Sekem's buildings.

Acknowledgements

We would like to thank our editors: Dr Bruno Sandkühler, Peter Segger, Rembert Biemond, Konstanze Abouleish and Christopher Nye. Konstanze Abouleish, Martina Dinkel and Regina Hanel contributed significantly to the creation of this book. We would also like to express our special thanks to Dr Bruno Sandkühler and Daniel Baumgartner for their support with the content, and to Samuel Knaus and Justus Harm for the photos. The initiative for the English translation came from Rembert Biemond, a member of the Sekem community. We thank Jeff Martin for the translation. For the English translation of the book, the content was, where relevant, updated from the German original to reflect the situation of Sekem in autumn 2023.

We would also like to thank our Sekem Partners all over the world for their invaluable support:

Culture

Alanus University
The German State of
  Baden-Wuerttemberg
Sekem Friends Germany
Sekem Friends Austria
Sekem Friends Netherlands
Sekem Friends Scandinavia
City of Stuttgart
Evidenz Foundation
World Social Initiative Forum
Witten/Herdecke University

Ecology

Biodynamic Federation,
  Demeter International
German Federal Ministry for
  Economic Cooperation and
  Development (BMS)
German Society for International
  Cooperation (GIS)
Greenpeace
IFOAM (International Federation
  of Organic Agriculture)
Agriculture Section at the
  Goetheanum

Economy

Alnatura
AUWA Holding
Organic Company
Davert
dm Drugstore
Cradle to Cradle e.V.
Ecor Naturasi
El Puente
GLS Bank
Tree of Life
Concolor AG
Austrian Development Bank
   (OeEB)
Odin
Oikocredit
Organic Flavour Company
Rapunzel
Triodos
Weleda
World Economic Forum

Social life

Artava AG
Home of Humanities
Right Livelihood Award
World Future Council
World Goetheanum Association

Notes

Preface
    1.   Abouleish, *Sekem*, p. 218.

2. Sources of Inspiration
    1.   Abouleish, *Sekem*, p. 218.
    2.   Ibid., pp. 47f.
    3.   Steiner, *The Spirit of the Waldorf School*, p. 98.
    4.   Abouleish, *Sekem*, p. 202.

3. The Sekem Symphony: From Initial Vision to the Present
    1.   Abouleish, *Sekem*, p. 33.
    2.   Ibid., p. 63.

5. Culture
    1.   Abouleish, *Sekem*, p. 187.
    2.   Steiner, *Truth and Knowledge*, p. 95.
    3.   *Sunan al-Tirmidhi* 2687, Book 41, No. 43.
    4.   *Abu Dawud* 3634, Book 25, No. 64
    5.   Surah 96.1–5.
    6.   Schiller, Fredrick, 'The Artists', in *Fidelo*, vol. 4 no. 1, Spring 1995, p. 55.
    7.   *Riyāḍ aṣ-Ṣāliḥīn*, No. 611.
    8.   Abouleish, *Sekem*, p. 194.
    9.   Ibid., p. 191.
    10.   Surah 95:4.
    11.   Steiner and Wegman, *Extending Practical Medicine*, pp. 11f.

6. Ecology
    1.   Abouleish, *Sekem*, p. 13.
    2.   Goethe, *Selected Poetry*.
    3.   Steiner, 'The Portal of Initiation' in *Four Mystery Dramas*, p. 17.
    4.   Eckermann, *Conversations with Goethe*.
    5.   Approximate figures are 600,000 trees and 500,000 tonnes of CO_2 in forty years.

6. According to the water standards established by the UN, a country is considered to be in water poverty if the annual water supply drops to less than 1,000 cubic metres of water per person. Egypt currently has around 550–80 cubic metres per person per year. See: Barakat, Shereif, 'Egypt Reaches Water Poverty Stage: Local Development Minister', *Egyptian Streets*, May 21, 2022, available at: https://egyptianstreets.com/2022/05/21/egypt-reached-water-poverty-stage-local-development-minister

7. See: Dr Eng. Ayman Elnaas, 'Verbesserung der Abfallwirtschaft in Ägypten' [Improving Waste Management in Egypt], *CDM Smith*, available at: https://www.cdmsmith.com/de/Client-Solutions/Projects/Solid-Waste-Management-Egypt

8. 'Data Page: Arable land use per person'. Our World in Data (2024). Data adapted from Food and Agriculture Organization of the United Nations (via World Bank). Retrieved from https://ourworldindata.org/grapher/arable-land-use-per-person

9. The project is funded and commissioned by the German Federal Ministry for Economic Cooperation and Development and the German Society for International Cooperation.

10. For its innovative approach in supporting farmers wanting to convert to biodynamic and other climate-friendly agricultural practices, Sekem was awarded in 2024 with the Gulbenkian Prize for Humanity by the former chancellor of Germany Angela Merkel, and recognised by the Environment Programme of the United Nations as a Champion of the Earth.

11. This does not include energy used in agriculture or transportation, only factors such as nitrogen from soils, biomass burning, fertiliser production, animal husbandry, or irrigation.

12. Intergovernmental Panel on Climate Change, *Climate Change 2014: Mitigation of Climate Change.*

7. The Economy

1. Abouleish, *Sekem*, p. 124.
2. Ibid.
3. al-Bukhari, Book 2, No. 13.
4. al-Bukhari 1434, Book 24, No. 37.
5. Steiner, *Rethinking Economics*, p. 63.
6. See *Food Wastage Footprint: Full-cost Accounting* from the Food and Agriculture Organization of the United Nations. Available at: https://www.researchgate.net/publication/337198849_Food_ Wastage_Footprint_Full-Cost_Accounting_Final_Report

8. Social Life

1. Abouleish, *Die Sekem – Symphonie*, p. 177.
2. Steiner, *Understanding Healing*, p. 87.
3. The inspiring words are all spoken in Arabic.
4. Steiner, 'At the Ringing of the Bells', in *Truth-Wrought Words*, p. 13.
5. *Faust: Part 1*, 'Night: Faust's Study (I)', 447–48.
6. Verse given to Edith Maryon on November 5, 1920. See Steiner, *Aufsätze über die Dreigliederung des sozialen Organismus und zur Zeitlage 1915 – 1921* [Essays on the Threefold Division of the Social Organism and the Current Situation 1915 – 1921].
7. Rumi, 'Only Breath', in *Selected Poems*, p. 32.
8. Quoted in Narain, *Myths of Composite Culture and Equality of Religions*, pp. 20f.
9. See Steiner, *Towards Social Renewal*.
10. al-Jayyousi, *Islam and Sustainable Development*, p. 78.
11. See, for example, Laloux, *Reinventing Organisations*; Robertson, *Holacracy*; Lessem and Schieffer, *Transformation Management*.

9. Our Vision for Egypt 2057 in the Context of Current Events

1. See Steiner, *A Psychology of Body, Soul and Spirit*, lecture of December 15, 1911.
2. See Gebser, *The Ever-Present Origin*.
3. See Steiner, *Theosophy*, pp. 38–47.

Resources

BOOKS

Abouleish, Ibrahim, *Die Sekem-Symphonie: Nachhaltige Entwicklung für Ägypten in weltweiter Vernetzung* [The Sekem Symphony: Sustainable Development for Egypt in a Global Network], Info3 Verlag, Germany 2017.

—, *Sekem: A Sustainable Community in the Egyptian Desert*, Floris Books, UK 2005.

al-Jayyousi, Odeh R., *Islam and Sustainable Development: New Worldviews*, Gower, UK 2012.

Beck, Don and Cowan, Christopher C., *Spiral Dynamics: Mastering Values, Leadership and Change*, Wiley-Blackwell, USA 2005.

Brotbeck, Stefan, *Zukunft: Aspekte eines Rätsels* [Future: Aspects of a Riddle], Verlag am Goetheanum, Switzerland 2005.

Eckermann, Johann Peter, *Conversations with Goethe*, Penguin, UK 2022.

Gebser, Jean, *The Ever-Present Origin*, Ohio University Press, USA 1985.

Goethe, Johann Wolfgang von, *Selected Poetry*, Penguin, UK 2005.

Khoury, Adel Theodor, *Der Koran – Arabisch-Deutsch: Übersetzung und wissenschaftlicher Kommentar von Adel Theodor Khoury* [The Koran – Arabic-German: Translation and Scientific Commentary by Adel Theodor Khoury], Gütersloher Verlagshaus, German 2001.

Laloux, Frederic, *Reinventing Organizations: A Guide to Creating Organizations Inspired by the Next Stage in Human Consciousness*, Nelson Parker, UK 2014.

Lessem, Ronni and Schieffer, Alexander, *Transformation Management: Towards the Integral Enterprise*, Routledge, UK 2016.

Narain, Harsh, *Myths of Composite Culture and Equality of Religions*, Voice of India, India 1990.

Robertson, Brian J., *Holacracy: The Revolutionary Management System that Abolishes Hierarchy*, Penguin, UK 2016.

Rumi, *Selected Poems*, Banks, Coleman (trans.), Penguin Books, UK 1995.

Scharmer, Otto, *Theory U: Leading from the Future as it Emerges*, Berrett-Koehler, USA 2016.

Steiner, Rudolf, *A Psychology of Body, Soul and Spirit: Anthroposophy, Psychology, Pneumatosophy* (CW115), SteinerBooks, USA 1999.

—, *Aufsätze über die Dreigliederung des sozialen Organismus und zur Zeitlage 1915–1921* [Essays on the Threefold Division of the Social Organism and the Current Situation 1915 – 1921] (GA 24), Steiner Verlag, Switzerland 1982.

—, *Four Mystery Dramas* (CW14), SteinerBooks, USA 2014.

—, *Rethinking Economics: Lectures and Seminars on World Economics* (CW340), Rudolf Steiner Press, UK 2015.

—, *The Spirit of the Waldorf School: Lectures Surrounding the Founding of the First Waldorf School Stuttgart – 1919* (CW297), Anthroposophic Press, USA 1995.

—, *Theosophy: An Introduction to the Spiritual Processes in Human Life and in the Cosmos* (CW9), Anthroposophic Press, USA 1994.

—, *Towards Social Renewal: Rethinking the Basis of Society* (CW23), Rudolf Steiner Press, UK 1999.

—, *Truth and Knowledge: Introduction to Philosophy of Spiritual Activity* (CW3), SteinerBooks, USA 1981.

—, *Truth-Wrought Words and Other Verses and Prose Passages* (CW40), SteinerBooks, USA 1979.

—, *Understanding Healing: Meditative Reflections on Deepening Medicine through Spiritual Science* (CW316), Rudolf Steiner Press, UK 2014.

Steiner, Rudolf and Wegman, Ita, *Extending Practical Medicine: Fundamental Principles Based on the Science of the Spirit* (CW27), Rudolf Steiner Press, UK 1996.

HADITH COLLECTIONS

al-Bukhari, Muhammad, *Sahih al-Bukhari* (9 Volumes), Arabic Virtual Translation Center.

an-Nawawi, Yahya ibn Sharaf, *Riyad as-Salihin*, Light Publishing.

al-Naysaburi, Muslim ibn al-Hajjaj, *Sahih Muslim* (7 Volumes), Dar-us-Salam Publications.

al-Sijistani, Abu Dawud, *Sunnan Abu Dawud* (7 Volumes), Dar-us-Salam Publications.

REPORTS

Intergovernmental Panel on Climate Change, *Climate Change 2014: Mitigation of Climate Change*. Available at: https://www.ipcc ch/report/ar5/wg3/

United Nations Food and Agricultural Organisation (2014). Food Wastage Footprint: Full Cost-Accounting. Available at: https://www.researchgate.net/publication/337198849_Food_ Wastage_Footprint_Full-Cost_Accounting_Final_Report

United Nation Environment Program (2019). Emissions Gap Report. Available at: https://www.unep.org/resources/emissions-gap-report-2019

United Nation Environment Program (2019). Global Environment Outlook 6. Available at: https://www.unep.org/resources/global environment-outlook-6

WEBSITES

www.cdmsmith.com/de/Client-Solutions/Projects/Solid-Waste Management-Egypt

Index

Abouleish, Ibrahim 7f, 18f, 28–34,
    44f, 50f, 54, 74, 77, 106f, 110, 129f,
    133, 136, 150f
agroforestry 98
animal husbandry 78f
associative economics 107
Aswan Dam 83
ATOS Pharma 112, 141

beauty, as path to the divine 51f
Beck, Don 24
biodiversity 97f
biodynamic agriculture 31, 61, 77,
    79, 86f, 102
— compost 81f, 117
— research trials 65

carbon emissions 101f
carbon sequestration 101f
climate neutrality 100–4
consciousness, development of 24f,
    152–55
— in the Quran 24
consciousness soul 153
conventional agriculture 77
Core Program 55f, 62, 72f, 147f
Cowan, Christopher 24
cultural life 43f
— in Egypt 58f

diversity and equality 139, 143f

EcoEnergy 95
EcoHealth 69

economy of love 106–11, 118–21
education (in Egypt) 57
Egyptian bee (*Apis mellifera lamarckii*)
    98f
Egyptian Biodynamic Association
    (EBDA) 32, 87
environment 25
eurythmy 52f

finance 122f

Gebser, Jean 152
Goethe, Johann Wolfgang von 23,
    76f, 80, 33
Graves, Clare W. 153
Greening the Desert (project) 88, 92

healthcare (in Egypt) 57
Heliopolis University for Sustainable
    Development 33, 56, 69, 86f, 95f,
    99, 102, 116f, 148
holistic research 64f
hygiene 53f

integrative health 67f
intellectual mind soul 152f
Intergovernmental Panel on Climate
    Change (IPCC) 101
Isis Organic 89, 112
Islam 20f, 27, 80, 110f

lifelong learning 46–51, 60–63

Nasser, President Gamal Abdel 139
NatureTex 28, 32, 141f
nutrition 77f

population growth (in Egypt) 138f
Prophet Muhammad 20f

Quran, the 22, 45f,
— and eurythmy 52f
— and nature 75f
— and the development of
  consciousness 24

recycling 115f
renewable energy 93–96
Right Livelihood Award 33
Rumi 134f

Scharmer, Otto 151f
Schiller, Friedrich 51
seeds 87
Sekem, beginnings 28–33
Sekem Future Council 7, 138, 145
Sekem Medical Centre 67
Sekem-Misa (currency) 124
Sekem Mother Farm 54f, 63, 82, 95,
  149

Sekem Sophia programme 141f
Sekem Vision Goals (an overview) 39f
sensitive crystallisation 66
sentient soul 152
social life, the 129–39
Spaces of Culture 71f
Spiral Dynamics 24, 141, 153
spiritual life 51f
Steiner, Rudolf 19, 24, 45, 54, 79f,
  107f, 132f, 152f
sustainable living 126f

Theory U 151f
threefold social order 25, 135
true cost accounting 87

ummah 131

Wahat El-Bahareyya (Wahat Farm)
  63, 82, 94f, 99, 103, 126f, 147f
Waldorf education 48
waste reduction 115f
wastewater treatment 92
water management 90f
water poverty 83f

zakāt 108–10

Praise for Sekem

'Sekem's vision combines the best of an ancient past and contemporary ecological knowledge. Sekem shows how abundance can be created in a desert and how ecology and economy can come together in an economy of love and care.'

– Dr Vandana Shiva, scientist and sustainability activist

'For over 40 years, Sekem has shown how the desert can be made fertile, how many people can work together, and how each person can be a bearer of the future through learning.'

– Ueli Hurter, co-leader of the Agricultural Section and member of the Executive Council at the Goetheanum, Switzerland

'For Egypt, Sekem offers the chance to gradually regain the pride and cultural significance that the country once experienced, but in the spirit of future-oriented, people-oriented, sustainable development.'

– Ulrich Walter, founder and managing director of Lebensbaum, Germany

'It's great to have a company in Egypt that has a vision for the future and is inspired by both the past and the present. In a wonderfully creative way, Sekem is thus brought to a special level of sustainable, entrepreneurial approach. This is what true institutional greatness looks like.'

– Prof Hossam Badrawi, chairman of the Nile Badrawi Foundation for Education & Development, Egypt

'Facing the challenges of our time, Sekem is a model for the future of sustainable food production and security, mitigation of climate change and a culture of dialogue.'

– Georg Stillfried, Austrian ambassador to Egypt

'The story of Sekem, how a desert was transformed into a place of human and planetary flourishing, is more relevant today than ever.'

– Otto Scharmer, senior lecturer at the Massachusetts Institute of Technology (MIT), and founding chair of the Presencing Institute

Floris Books

For news on all our **latest books,**
and to receive **exclusive discounts,**
join our mailing list at:

florisbooks.co.uk/signup

Plus subscribers get a FREE book
with every online order!

We will never pass your details to anyone else.

www.ingramcontent.com/pod-product-compliance
Lightning Source LLC
Chambersburg PA
CBHW051258020426
42333CB00026B/3255